Marianne Landzettel

Regnerative Agriculture
Farming with Benefits

Profitable Farms.
Healthy Food.
Greener Planet.

Farming with benefits

Regenerative Agriculture
Farming with Benefits
Profitable Farms. Healthy Food. Greener Planet

Copyright: Marianne Landzettel 2021

All rights reserved. No part of this publication may be reproduced, stored in a retrieval system, transmitted in any form or by any means electronic, mechanical, including photocopying, recording or otherwise without prior written consent of the copyright holders.

Published by World University Service, Wiesbaden, Germany

ISBN13 978-1-8383805-0-2
GTiN 9783922845515

Design and artwork by DM Design & Print

For Willy Dietzmann, agricultural extension agent and my grandpa. He was passionate about food and farming and full of love for his granddaughter.

I still miss him.

Contents

Introduction by Nicolette Hahn Niman	v
Prologue: Apocalypse Now and what to do about it	vii
1. Good news from the underground	1
2. Farming with nature and building communities	17
3. The 'Dust Bowl' – a history lesson not learnt	37
4. Hawai'i – Industrial agriculture's ground zero	49
5. Ahupua'a – Hawaiians relearn growing their own food	83
6. Profits on the prairie	109
7. Farming the margin	127
8. Farm futures	151
9. "Don't farm naked" – plant cover crops	177
10. Kick-starting the system	201
11. Finding new paths	215
12. "We've maybe got ten years"	235
Acknowledgement	255
Index	257

INTRODUCTION

Here in the United States, many farmers and ranchers have long been skeptical of climate change. In 2015, after I gave a talk in Missouri about sustainable livestock farming, a farmer in his sixties approached me and said he really enjoyed my speech. But, he followed up, "Why do you talk about global warming like it's real"? It was a jarring moment. As a person trained in biology and law, I have always tried to seek knowledge and truth, looking hard at credible, empirical scientific evidence. At that moment I was face-to-face with the reality there were people in rural America whose views on climate change were fixed, and they were doubters. Many were listening to elected officials and news sources who were repeatedly suggesting this was "fake news." And in their own lives, they were not seeing evidence of a dramatically changing climate.

I believe that's changed over the past several years. Coastal parts of the United States have seen a dramatic rise in the number and severity of hurricanes. The Western U.S. has been experiencing longer and more extreme wildfire seasons. Even the center of the country – "The Heartland" – has been experiencing newly extreme storms and flooding. Facebook has been filled with pictures of Biblical flooding in Nebraska and wind twisted silos and barn roofs in Iowa. The physical evidence of climate events directly affecting rural America has jostled loose the once unshakable view that climate change was merely a "hoax." "As the extreme weather that caused flooding in the Midwest [in 2019] dovetailed with the scientific warnings, climate change and its impact on agriculture started to get more attention—including among farmers, lawmakers and farm groups that have been reluctant to connect the dots," *Inside Climate* News reported.

At the same time, an idea has been taking hold in mainstream agriculture. That idea is regenerative farming. American farms and ranches have struggled financially for decades. A growing number have been moving away from the once-ubiquitous mindset that agricultural chemicals are the solution. More are turning, instead, toward farming that focuses on soil health, and seeks to minimize chemical inputs. This doesn't get a lot of mainstream media attention. But I have seen the signs everywhere. I hear it when I speak directly with people in agriculture. I see it in social media conversations among farmers. A university professor friend of mine who writes and consults about regenerative farming told me his conference talks – once attended by handfuls – are now crowded with hundreds of farmers eager to learn of a different, better way. This is not the "hippie farmers" of the sixties; it's mainstream agriculture.

But this movement is not being led by academics or government, at least not in the United States. It's being led by farmers. A couple of years ago, I had the distinct pleasure of spending time with one of the heroes of this burgeoning movement, Gabe Brown. In his book *Dirt to Soil*, Gabe describes his own journey from conventional North Dakota crop farmer to the architect of an ecologically complex, highly diverse, regenerative farm. As I sat next to Gabe at the authors' table in Kentucky, a long line of farmers waited patiently to buy his book and get his signature. Gabe would not call himself an "expert" of sustainable farming. But he is inspiring thousands with his example. He has shown, through experimentation and creativity on his own farm, the way to make an ecologically vibrant and financially viable farm. One that produces delicious, nutrient rich food. His example is a beacon of hope for farmers around the country, and even in other parts of the world. Getting to know Gabe and other farmers who are showing the way forward for regenerative farming has given me great hope for the future.

I am delighted Marianne Landzettel has written this book, sharing the stories and examples of regenerative farmers and farms. We need regenerative agriculture to address the urgent (and related) ecological problems of climate change, soil erosion, water scarcity, drought and desertification. And we need it just as much to address the health crisis of diet related chronic diseases. By telling the stories of regenerative farmers and ranchers, Marianne shows us what is possible. Farmers and consumers, together, can forge a new and much better future for farming, food production and human health.

Nicolette Hahn Niman

Rancher, Lawyer, Author, Mother.

Bolinas, California

October, 2020

Apocalypse Now and what to do about it

Lunch time in Chico on 8th November, the day the Camp Fire broke out

Chico is a small university town in northern California. Cynthia Daley teaches there and is the director of the Regenerative Agriculture Initiative. We had come to California at the end of a long research trip that had us travelling through Kansas, Nebraska and Colorado. It was the second week of November, and Cynthia Daley had agreed to meet us just south of Chico on the university's research farm. But then things changed, and they did so rather fast. In the early hours of that day the 'Camp Fire' broke out. While the small town of Paradise burnt to the ground, people fled for their lives and the flames spread, Chico grew dark and cold under the clouds of smoke that fanned out as far as the Pacific. In Chico, just over ten miles to the west of where the fires raged, ash particles silently floated to the ground like snowflakes. Watching TV news and the continuously updated maps, we began to understand the scale and extent of the fire. We stood outside, looking at the red glow on the eastern horizon under an otherwise dark sky. It felt as if nature was posting some final climate change warning: in November, the rainy season should have been in full swing, there should have been no fires, California should have been green. But since the 1970s, climate change has led to the fire season being prolonged by more than 80 days.

The morning of 8th November, just west of Chico

Climate change is creating a flood of negative headlines on a daily basis. But amidst the gloom and doom there is one bit of seriously good news: regenerative agriculture[1] can sequester carbon in the soil and thereby reduce the CO_2 concentration in the air. We know what regenerative agricultural practices are, and they are available to be employed immediately. While not every practice is suitable for all farms, every farmer will find some regenerative practices that will be right for his or her land. Farmers who work with regenerative agriculture not only help to mitigate climate change but can potentially contribute to actually cooling the planet. And carbon rich 'good' soil can also hold more water, which helps farmers to cope better with drought conditions, preventing run-off, and in

---

1   Regenerative agriculture is not a standardised term but rather a collection of practices. Certified organic agriculture can be a form of regenerative agriculture but there are exemptions. For this book I am using the term as defined by the NGO Carbon Underground: Regenerative Agriculture is a holistic land management practice that leverages the power of photosynthesis in plants to close the carbon cycle and build soil health, crop resilience and nutrient density. Regenerative agriculture improves soil health, primarily through the practices that increase soil organic matter. This not only aids in the increase of soil biota diversity and health, but increases biodiversity both above and below the soil surface, while at the same time raising both water-holding capacity and sequestering carbon at greater depths. This draws down climate-damaging levels of atmospheric CO_2, and improves soil structure to reverse civilization-threatening human-caused soil loss.
https://secureservercdn.net/184.168.47.225/02f.e55.myftpupload.com/wp-content/uploads/2017/02/Regen-Ag-Definition-7.27.17-1.pdf

some areas, protecting low-lying towns and cities from flooding.

Why is this book mostly about regenerative agriculture in the United States of America? In Europe, agriculture has evolved over thousands of years. Agriculture is not just a function of geography, soil type and climate, it's also shaped by regional traditions, history, politics and policies. Agriculture in the Netherlands and northern Germany differ even though geographical and climatic conditions are very similar. Agriculture in east Germany is different from that in west Germany because the large communist farming cooperatives that were created in the GDR lent themselves to industrial farming practices after the fall of the Iron Curtain.

Such variations in conditions and differences in development can obscure the view. In the US climatic and geographic conditions are much more homogenous: from the east of Kansas to the west it's about as far as from Hamburg to Munich, or from London to the north of Scotland but while landscape and agriculture will change continuously as you drive through Germany or from the south of England to the north of Scotland you will see fields of corn, soy, sorghum and wheat just about everywhere in Kansas. Only the number of center pivot irrigation systems will increase the further west you are; average rainfall in eastern Kansas is considerably higher than in the west.

And large parts of the United States have only been farmed for about 150 years. In some regions, agriculture even has a start date: May 20th 1862, the day President Abraham Lincoln signed the 'Homestead Act'. The bill allowed settlers to claim 160 acres of land. A settler could become official owner of a piece of land after five years if he could prove that he had continuously lived on it and farmed it. For many new immigrants, the 'Homestead Act' seemed like a golden opportunity. Settlers streamed to the more remote parts of the Midwest and the Great Plains, and plowed up the prairies.

The third reason to look towards the USA is the fact that everything there is simply bigger – including the problems. The consequences of climate change can be seen the world over, but in the US they are now impossible to overlook. Drought, flooding, and blizzards, weather events that year after year break all records for highest or lowest temperatures ever measured. Farmers in the US don't just see the effects of climate change on their farms, they are forced to deal with them and find solutions.

In February of 2019 NASA scientists announced that in 2018 the average surface temperature on Earth had been the fourth highest on record. Newspapers and electronic media outlets illustrated the news with graphs and maps showing much of the globe in hues of yellow to red.

There is no other news today

Today, nature writes headlines: forest fires, floods, and hurricanes cannot be ignored.

We know that time is running out to stop the atmosphere from heating up even more. That means cutting Greenhouse Gas (GHG) emissions, with reducing the use of fossil fuels top of the list. The manufacturing industry, the transport sector, and construction are responsible for most of the emissions. But industrial agriculture plays a major role, too: through direct emissions from raising livestock in confinement, to agricultural practices such as plowing, and indirectly through the use of agrichemicals, to name but a few of the elements.

A lot of technical fixes are being bandied about, from huge plants 'cleaning' the atmosphere to spraying tiny sulfate particulates into the lower stratosphere. These are solutions that would cost billions and are nowhere near ready to be rolled out on a scale that would have any effect.

So, is it 'solution found, agriculture to the rescue'? That is pretty much the idea behind the '4 per 1,000' initiative which was founded in France at the 2015 UN Climate Change Conference under the auspices of the then French Minister of Agriculture, Stéphane Le Foll: 'An annual growth rate of 0.4% in the soil carbon stocks, or 4‰ per year, would halt the increase in the CO2 concentration in the atmosphere related to human activities,' states the initiative's website[2]. And: 'The "4 per 1,000" initiative shows that agriculture can provide some practical solutions to the challenge posed by climate changes, while at the same time highlighting the challenge of food security through the implementation of agricultural practices adapted to local conditions such as: agroecology, agroforestry, conservation, agriculture, landscape management....' Several governments signed up: the Germans, the Spanish, and, in 2018, the Vietnamese, as did universities and other research bodies from around the globe. Numerous studies are being conducted, and the conferences and meetings continue. The international connections that are being forged no doubt help the flow of information on a scientific as well as a practical level. What hasn't happened (yet) is a government coming forward with a practical plan that puts money behind the idea of carbon sequestration through regenerative agriculture, and with proposals on how to remunerate farmers for this service.

With the Soil Health initiative, California has taken a tentative first step which will be described in the last chapter of this book.

New ideas are emerging and gaining momentum. In February, a blueprint for a Green New Deal[3] was published, 'urging a '10-year national mobilization' for a speedy shift away from fossil fuels and calling for national health care coverage and job guarantees in a sweeping bid to remake the U.S. economy'. The scope is broad and the time frame ambitious, given the glacial pace at which such political processes move.

The European Green Deal is the roadmap set out by the European Commission with the aim of making the countries in the EU climate neutral by 2050.

Politicians in Britain, too, are under pressure to act. At the end of April 2019 tens of thousands of people in London and other British cities came out every day for a whole week to protest. Their demand was for politicians to declare a climate and ecological emergency and GHG emissions to be reduced to zero by 2025. These goals may not be achievable, but one thing is clear: a sizable number of people have understood what is at stake and are demanding action. Many of the protesters are young. The 'School strike for climate', initiated by the Swedish student Greta Thunberg, has developed into a global 'Fridays for Future' movement which has students staying away from school and demonstrating instead. Politicians in Europe are taking note. Whether they will take action remains to be seen.

---

2   https://www.4p1000.org/
3   https://www.politico.com/story/2019/02/07/green-new-deal-resolution-1155146

What we can do...

'We've got maybe ten years to fix this,' said Cynthia Daley, when we talked at the university farm the day after the 'Camp fire' started. With that time frame, hoping for political solutions may not be an option. But it could lead to a completely different approach: a change of perspective.

What if we focused on farm profitability with carbon sequestration as a welcome and highly beneficial side effect?

Noon the next day at the Chico University Farm

Here's how I believe different and seemingly independent developments may align to make this an option. Industrial farming is in deep crisis. Farm net income in the US is at its lowest level in 15 years. Commodity prices are falling while input costs, in particular for seeds, herbicides and fertilizer, are rising. Farmers are dealing with increasing herbicide resistance in weeds for which, this time, the agrochemical industry does not seem to have another poison fix to offer. There is no hint of news that a new herbicide might become available soon; what the industry is offering is new combinations of already available herbicides which may be efficient for a short while. However, in June 2018, scientists in Missouri confirmed the first case of a waterhemp plant that had developed resistance to six different herbicides[4]. And of course, the herbicide combinations can only be applied on soy and corn plants that have been genetically engineered to withstand them. Developing them doesn't come cheap, which again increases input costs for farmers.

Then there are the consumers who remain wary of GMOs and are becoming more alert to possible links between pesticide use and human health. Demand for organic food, not just in the US but worldwide, is increasing. Consumers are also starting to ask questions about the use of antibiotics and growth hormones in milk and meat production, animal welfare standards, and the impact of industrial farming practices on the environment. People with such concerns are still a minority, but it's a vocal minority and a younger demographic. In 2020, people who normally wouldn't take much interest are considering such issues too – in the midst of the Coronavirus crisis, perspectives change: food, where it comes from, and how it is produced has become an important and urgent topic.

As a commodity farmer or meat or dairy producer, you can of course try to produce more in order to survive. And you can hope for better prices, and for the agrochemical industry to come up with some magical fix to fight the weeds, increase the milk yield, or make your beef cattle pile on the pounds with less feed. And as a last resort, there is always the farm bill and crop insurance. Or you do what so many farmers were forced to do during the last US farming crisis in the 1980s: you sell the farm or lease the land to your neighbors and move away in search of work.

But there are farmers in the US who take a radically different approach: they work on ways to reduce their input costs and achieve a better margin. And that's where healthy soil comes into the equation. Carbon-rich soil teeming with soil life, from earthworms to microbes and fungi, will be so fertile that no or very little chemical fertilizer is needed. These farmers will use whatever regenerative agricultural practice they can in order to get their soil into optimal condition. They do so because it makes their farm (more) profitable. As a side effect, the carbon rich soil they are creating benefits the environment by reducing the CO_2 level in the atmosphere.

It's called farming with nature. To farm this way makes sense. It's a win-win situation. So why is everyone not doing it? Because it's difficult and farmers who try are on their own. Literally. There may be no other farmer in the same or even the next county who works with regenerative agricultural practices. Soil,

---

4 http://wssa.net/2018/06/scientists-confirm-first-case-of-waterhemp-with-six-way-resistance/

climate, elevation, average rainfall – all these factors will likely vary. Each farmer will face unique problems and will have to find his or her own solutions. And the problems don't stop at the farm gate – there is market access, infrastructure, sourcing seeds, tools, and machinery – how to deal with each of these issues is for the individual farmer to work out. It takes real resolve to do all that in a rural community under scrutiny from the conventionally farming neighbors, those who are likely to discuss every mishap and failure at length in the coffee shop.

This book attempts to make the case that regenerative agriculture can help farms to become and stay profitable while delivering climate benefits as an advantageous 'side effect'. And the following chapters will look into what we can do to help and support farmers. We all eat food several times a day, and with the food, we buy and the meals we cook we make choices: we can either support organic and regenerative farmers or food conglomerates and the agrochemical industry.

Regenerative agriculture isn't one thing. What practices a farmer can employ, what works, what might work and what probably won't work depends on a huge number of factors, including climate, longitude and latitude, elevation, and the farm's history – improvements may be much slower on very degraded soils that have been intensively farmed for a long time.

With that in mind, the research for this book turned into a number of road trips, with my husband, Martin, doing most of the driving and taking pictures on the farms while I kept asking yet more questions.

We started out in Hawai'i to see the sites not featured in any travel guide. Agrochemical companies operate their test sites away from palm-fringed beaches and fancy hotels. These test plots are essential for the development and maintenance of parent lines for hybrid and genetically engineered seed varieties and involve the use of extraordinary amounts of herbicides and pesticides. Hawai'i is like ground zero for anyone wanting to observe the effects of excessive, long-term use of agrochemicals on soil, water, and air quality.

The Hawaiian Islands are rather small, with steady trade winds. Since the agrochemical companies came to Hawai'i in the 1990s many communities are being exposed to drift and pesticide-laden dust. While it may not be possible to directly link exposure to health problems occurring at some later date, it is possible to say that health professionals see some diseases – like asthma, skin problems, and rare cancers – at a significantly higher than average rate. Hawaiians[5] have fought hard for legislative change, and in 2018 regulations were passed that give better protection to schools and allow better access to information. The anti-pesticide campaign also revived interest in the ancient Hawaiian farming system that provided food at a time when importing perishable goods was not an option. Today, about 80% of food is shipped from the US mainland. But more Hawaiians are growing their own food, and some run successful, profitable farms while training and educating a new generation of young farmers.

Regenerative agriculture gathers traction, not just in Nebraska

Our next trip took us to the Rodale Institute in Pennsylvania, and an organic grain farm in upstate New York. At Rodale, a long-running comparative field trial has established the benefits of organic and regenerative agricultural practices for the soil. The techniques are put to the test on an organic grain farm in northern New York. 'Every field tells a story,' says Klaas Martens, who runs the farm. Keen observation and adjusting farming methods accordingly has a lot to do with the commercial success of the farm. The organic feed and seed business operated by his wife, Mary Howell, is another element.

---

5   Sometimes 'Hawaiian' is used to describe people of Polynesian heritage whose ancestors first settled on the islands. I am referring to citizens of the State of Hawaii, whatever their ethnic origin may be.

Before settlers turned the Midwest and the High Plains into the agricultural 'heartland', most of the region was covered by native prairie grasses. Wet grass prairies in Iowa, tall grass prairies as far to the east as Ohio, short grass prairies as far to the west as Colorado and Texas in the south. Since these prairies were plowed into farmland a lot of topsoil has been degraded or lost altogether. Some farmers are trying to stem the tide. They use regenerative agricultural practices to rebuild their soils and their businesses.

In fall of 2018, we drove through Colorado, Nebraska and Kansas to visit some of these farmers. Rainfall – or the lack of it – proved to be a decisive factor: regenerative farming is different in a dry area; a higher average rainfall gives farmers a lot more options. But all the farmers and ranchers we visited ran diverse and profitable businesses: from cattle raised on short grass and mixed grass prairies to growing wheat, corn and soy. Some of the farmers work with cover crops, produce seeds for other growers, and run a seed business. They market directly or through a cooperative, they sell online – through brokers or through an auction. They produce to fulfill individual contracts or do their own processing. All of them assess and evaluate their options and look for new ones: they think about branding, adding more value on the farm, diversifying, scaling down, and stacking businesses. And they see themselves as part of a regenerative farming community. Each farmer has to take decisions that work under the particular circumstances of their farm, but they are eager to share their experiences, their failures as well as their successes. What works on one farm may not work on the next, but it might spark an idea that eventually leads to a solution. 'Diversity' was a term that came up in every conversation and in different contexts – from biodiversity to diversity of farm animals, from companion cropping to 'crazy fields', from diverse farm businesses to diversification in marketing.

The last leg of our trip brought us to California. Farmers in the Central Valley produce the majority of all fruit and vegetables consumed in the US. Nuts are another important high value crop – walnuts, pistachios and almonds; 80% to 90% of the world's almonds come from California. The state has extremely fertile agricultural land, but its southern part in particular has experienced long periods of drought. The most recent one lasted six years – from 2011 to the winter of 2017/18.

Fruit and nut growers are particularly hard hit by climate change. The winter months are not as cold anymore, and are interspersed with warm periods, meaning trees don't go into dormancy, the rest period they need to stay healthy and productive. We visited a walnut grower who is developing new ways of farming and improving soil health to keep his orchards profitable and in production.

The northern part of the Central Valley is well known for its rice farms. The Lundberg Family Farm is the largest organic rice producer in the US, with its own lab and nursery. How to grow organic rice successfully and profitably is something the Lundbergs had to work out for themselves, including breeding new rice varieties that are suitable for the local climate, as California differs from most other rice-growing areas.

We met Bryce Lundberg on a beautiful and warm day in early November. It was nearly the end of our trip. We took our time driving back to Chico, stopping to watch the huge flocks of migratory birds settling on the wetlands at dusk. The next morning, the 'Camp Fire' started. With roads blocked and traffic around Chico at a standstill, the meeting with Cynthia Daley at the CSU Chico Farm was postponed till the following day. By then our conversation took on a sense of urgency, heightened by the surreal but literal experience of 'darkness at noon'.

There is no escaping the fact that 'nature' is a finely balanced system. Humans are part of it and everything we do has consequences within this system, good and bad, intended and unintended. Improving soil health might be one of the rare win-win moves we can make. Regenerative farming practices can be employed with the intention of maintaining or increasing farm profitability while having a beneficial effect on the climate and the environment, too.

The connection between regenerative agriculture, soil quality, environment, climate and profitability is not just important for farmers, it matters for all of us. If we understand the system, if we know how our food is produced, and if we actively support regenerative farmers, we can mitigate climate change a tiny bit with every bite we eat. 'Eating is a political act'[6], said journalist and author Michael Pollan in a discussion about his book 'An omnivore's dilemma'. With regenerative agriculture, we can turn eating into an act of environmental protection and an active contribution to climate change mitigation.

This book is about flipping the perspective and farming with benefits: for profitable farms, healthy food and a greener planet.

Initially I intended to write this book only in my native tongue, German. But the weeks and months spent writing took me back to the US, the farms we visited, the conversations we had, and I thought: I owe it to the farmers to tell their stories in English, too. I started translating and rewriting the German version in the fall of 2019. Just a few weeks in the devastating wildfires in Australia broke

---

6   https://www.cornucopia.org/2008/11/michael-pollan-eating-is-a-political-act/

out. The story long since has been a global one. The struggle of regenerative agriculture, keeping the farm profitable and dealing with the challenges of the climate crisis concerns us all.

Good news from the underground

Earthworm heaven

'It is so much more complex than we could ever imagine,' says soil microbiologist Kris Nichols. 'When I was an undergrad student, we thought we knew about ten percent of soil organisms. Today we think it's maybe one percent.'

Nichols doesn't really have time to talk; she's in the middle of moving. Until recently, she was responsible for soil research at the Rodale Institute[1], but now she has decided to return to the Midwest, where she grew up. For her, soil biology is far more than an interesting research topic. Healthy soil with a functioning soil biome is the basis of life; literally, it is essential for plant growth, for healthy crops, healthy food, and the future of the planet, she says. More people need to know about these connections, and not just farmers, but anyone who eats.

Nichols is a scientist, but also a practical farm consultant and educator. She wants to work with anyone willing to engage with what turns out to be a fascinating but complex topic: nothing in nature is linear, everything is part of a cycle, a system and a network – and 'much more complex than we could ever imagine'.

---

1   More about the Rodale Institute in Chapter 2.

Soil biologist Dr. Kris Nichols

When Kris Nichols talks about soil biology, it's like the visit to a planetarium, but in reverse: while astronomers use gigantic telescopes to discover new galaxies, soil biologists like Nichols peer through electronic microscopes that reveal to us the otherwise invisible world underground.

It's a complex world with infrastructure and transport systems where pretty much everything is available – at a price in the right kind of currency. The most commonly accepted 'payments' are the simple sugars which only plants can produce. Through the process of photosynthesis, they are able to use sunlight in order to convert carbon dioxide and water into simple sugars. The plants absorb carbon dioxide from the air and release oxygen – this is why green spaces, particularly in cities, are so important: plants clean the air.

From its leaves the plant sends some of the sugars down to its root system, where they can be 'traded' for pretty much everything needed for growth and more photosynthesis: water, different minerals, and trace elements like nitrate, phosphorus and potassium, but also calcium, sulfur, zinc, magnesium, iron, and copper. Some of these minerals may be within direct reach of the roots but remain inaccessible because they haven't been 'processed' or are not part of a more complex molecule that the plant can utilize. And this is where soil organisms come in: bacteria and algae, fungi, worms, insects, arthropods such as millipedes, nematodes, and protozoa. All of them have a job to do. One teaspoon of soil can contain five billion bacteria, 20 million fungi and one million protozoa and algae. A functioning soil ecosystem will be densely populated. Of course, not every organism is there to contribute to photosynthesis and the wellbeing of the plants we see above ground. Each has a life of its own. It feeds, excretes, eats, and gets eaten, and thereby fulfills some function within the (soil) ecosystem. The division between a world above ground and another one below is rather arbitrary: it's not just plants that have roots in the ground and other, mostly green parts visible, such as leaves, blades of grass, stalks, branches, and needles, above ground. Earthworms 'import' food by dragging rotting organic substances from the surface deep into the soil. Other soil organisms then transform this organic plant or animal material into soil organic matter (SOM), something you'll hear a lot more about in this book. A high percentage of soil organic matter is an important numerical indicator for a well-functioning soil ecosystem. A second indicator is biodiversity: the more species there are, the more different and specialized roles they can fulfil, each contributing to the system running more smoothly and efficiently. And this isn't just true for soil, but ecosystems in general: the biggest problems arise in monocultures. With chemical fertilizer, farmers can cover the plants' basic nutritional needs, but monocultures also turn out to be like a land of milk and honey for particular pests. In a biodiverse ecosystem, there will be beneficial insects ready to keep the pests in check. A monoculture doesn't provide good conditions for insect diversity, including the beneficial insects. As a result, their numbers remain low while the pests thrive in a habitat that suits them perfectly. That's why in order to save their canola, corn, or wheat harvests, farmers often have no choice but to use pesticides.

Urbanites, too, are starting to understand the damaging impact of monocultures: nowadays, bees kept in cities have a better chance of survival than their cousins in the countryside. Parks, gardens, and street trees provide more biodiversity and better sources of nectar throughout the seasons than, for example, a canola field. Once the flowering period is over, bees and other pollinating insects often struggle to find a food source unless the farmer increases biodiversity by planting wild flower strips or flowering hedges.

But back to the soil ecosystem. If lots of different plants grow together, as they would for example in a flowering strip, they often benefit from each other's presence – a plant with broad leaves is able to catch a lot of sunlight while at the same time providing protection for a plant that prefers shade and cooler conditions. And in the long run, the diversity above ground enhances diversity in the soil which, once again, benefits plant growth. The higher the diversity of plant and animal species, the better the chances that somewhere in this complex system the right nutrients and habitat can be found. The decisive factor is balance: in a functioning, highly diverse ecosystem, no one species dominates.

Organic substances make up only a small part of what is considered to be fertile soil. A little less than half consists of minerals which include sand, loam, and iron oxide. These inorganic substances have different particle sizes, which makes it easy to distinguish between them. All you need is a bucket of water: large particles like gravel, sand, and silt sink quickly to the bottom, and the water remains clear. Loam particles are so small that they seem to float in the water, turning it murky and brown.

The other half of good soil consists of air and water – soil organisms need both for survival. For soil to be able to hold water like a sponge and air like a soufflé, it has to have structure.

Good soil smells good

The glue that holds the particles together which in turn form the soil structure is called glomalin – a discovery that was only made in the 1990s.

What exactly glomalin is and how it glues together soil particles so that both water and air can be stored is fascinating but also quite complicated.

Glomalin is produced by mycorrhizal fungi in the soil. About 80 percent of all plant species live in symbiosis with mycorrhizal fungi. Often, the fungi penetrate the plant's roots, which allows a continuous and seamless exchange: the plant makes simple sugars available to the mycorrhizal fungi, which reciprocate with nutrients and water molecules. Mycorrhizae form hyphae, long, ultrathin 'extensions' which form a vast network and are therefore able to reach not just further but also deeper into soil layers that would otherwise be inaccessible to the plant. They absorb water and nutrients which are transported back to the plant roots through the 'pipeline system' of the hyphae network. In 'exchange' for the water and nutrient deliveries, the fungal network receives sugars. Mycorrhiza and their complex role in the soil ecosystem is a topic that Kris Nichols has particularly focused on in her research.

Glomalin is a double-sided coating on the hyphae, she tells me. It is extremely important because it allows the hyphae to go through the air-water interface in the soil and withstand the difference in pressure between air and water molecules. Different fungi produce a lot of different types of coatings. Some of them are hydrophobic, or water-repellent. Glomalin is a glycoprotein, which means that it has sugar groups associated with it which are sticky. And the amino acid structure that is part of the protein can be more hydrophobic. Basically, one can imagine glomalin like a sticky tape, sticky on one side and water-resistant on the other. When the hyphae degrade, the glomalin sloughs off. 'One of the things we found is that as it attaches to other soil particles or attaches to aggregates, it helps those aggregates to be water stable. The sticky side is on the inside of the aggregate and helps to glue it together, but the water insoluble side is on the outside. It puts a kind of waxy coating on the outside,' Nichols explains. Why that is important becomes evident when it starts to rain. 'When the soil aggregate is dry, there is open space inside the aggregate; it's filled with air. When it rains water moves into the soil, it moves inside the aggregate, and it fills up the space where the air molecules are. But air, as it is less dense, moves at a slower rate than water does. So, the water moves into that open space, but the air molecules don't move out fast enough and they get condensed and pushed closer and closer together, pressure builds up, and eventually, the aggregate blows up, exploding from the inside out. If some of the space on the surface is sealed off with the glomalin or these waxy substances, there is less space so that the water comes in at a slower rate. That allows time for the air molecules to get out or to dissolve in the water. And now the aggregate can fill with water without blowing up.'

Thanks to the glomalin 'glue', soil keeps its structure, and rain water can not only infiltrate the soil, it can also be stored. The difference is very important: when it rains, in particular during the torrential downpours which are increasing in frequency, the soil quality is the decisive factor for what happens next. The soil in conventionally managed fields is often compacted with little soil life and the water infiltration rate is therefore low. Most of the water will run off such a field as if it were a table top, taking soil particles with it. During a torrential rain event several tons of topsoil can be washed off. Depending on the time of year, seeds or seedlings can be swept away, too.

Soils with lots of soil organisms and a functioning soil ecosystem, on the other hand, act like a huge sponge which absorbs water until all its pores are full to capacity. Only then does water start to drip from the sponge. Applied to soil, this means: the thicker the humus layer (the sponge), the more water can be stored, and the less runs off unused. And that is very important with regard to climate change: water stored in the soil remains available to plants over a long period. During a drought, such water resources can be decisive for the survival of the plants – and, figuratively speaking, for that of the farmer, too, because he will either have a harvest or he won't.

Plants also help to increase the water infiltration rate. During torrential rain events raindrops hit the ground with high velocity and within a short period, the surface can turn into mud, making water infiltration harder. If the ground is covered by plants, their leaves will counter the velocity of the water drops, and the rain can infiltrate the soil. That is one of the reasons why cover crops (about which you will read a lot in the coming chapters) are of such importance.

Glomalin is important for yet another reason: it is a source of 'stable[2]' carbon which will be stored in the soil long term. CO2 is one of the greenhouse gasses causing climate change. One method to reduce the level of CO2 in the air and mitigate the climate crisis is to get carbon back into the soil – permanently. Put differently: the more glomalin the mycorrhizal network produces, the better it is for the climate and for farmers. Plants which are supplied with the optimal amount of nutrients and water will yield more. Kris Nichols explains why that is the case: 'Fungi will build glomalin in a relatively short time period. This becomes an issue if there isn't enough carbon to do it. In order to get carbon, the fungus has to satisfy the needs of the plant. It is a supply and demand issue. The demand is most often: give me nutrients, I'll give you more carbon.' The fungi

---

2   There is also 'labile' carbon. Within the carbon cycle, it is initially stored and then released again. Plants absorb CO_2 from the air via photosynthesis; when the plant dies and decays, CO_2 is released again.

extend the hyphae network to supply the plants more efficiently and thereby get hold of enough carbon. Once the system works well the mycorrhiza will have a surplus of carbon available which can be used to produce glomalin. The glomalin layer stabilizes and protects the hyphae and contributes to optimizing the system further. That's why tillage is a real problem in agriculture, says Kris Nichols: 'Tilling destroys the hyphae network, it destroys the mycorrhizae. As a result, the fungi continuously have to repair the hyphae, and there is not enough surplus carbon available to build glomalin.'

No-till agriculture is one method to sequester carbon and store it in the soil long-term. Unfortunately, most crops – cereals, potatoes and vegetables – are annual plants that have to be seeded and planted anew each year. Growing seasons are usually short, and in order to get healthy plants and high yields, farmers try to give these annuals optimal conditions from the start by plowing and preparing a 'seedbed'. Seeding directly into a layer of vegetation covering a field is difficult and sometimes completely impossible. And in a field where crops struggle because there is too much 'weed pressure' – competition for nutrients, light and water from weeds – yields will be down to the point of being a complete economic loss. But farmers have developed methods to disturb the soil as little as possible and remain profitable – more about that in the next chapters.

Only one form of agricultural production needs no plow and achieves both – to produce food and sequester above-average amounts of carbon below ground. That production system is keeping grazing animals on permanent grassland.

The key term here is '*permanent grassland*'. Vegans are correct when they argue that consuming meat and dairy products from intensively raised animals contributes to climate change. With intensive animal agriculture it takes 7kg (more than 15 pounds) of corn or cereals to produce one kilogram of beef. The result isn't just a massive carbon footprint, but considering that the world population continues to increase, we cannot afford to grow animal feed where growing food for humans is an alternative. The arguments vegans make are correct with regard to *intensive* animal agriculture but totally wrong when talking about grazing animals on permanent grassland. Permanent grassland is not in competition with cropland; usually it can only be found where other crops can't be grown. We're talking about areas that are too dry, too wet, or too hilly to plant cereals or corn.

Ruminants like cattle are the only creatures that can live off grass because they have not one, but four stomachs which enable them to digest it. Therefore, only ruminants are able to convert solar energy into calories and, eventually, meat and dairy, food we humans can eat. Of course, the existing grasslands can sustain only a relatively small number of animals. The globe isn't big enough to keep on grass all the world's animals currently being kept in confinement and fed with cereals and soy. Our present levels of meat and dairy consumption are totally unsustainable and we have to reduce them. But the meat and dairy products we do consume should be produced by grass-fed animals.

For everyone to go vegan is not the answer – animals are an integral part of sustainable agriculture. And cattle on grassland benefit the climate: ruminants and grasslands evolved together over millennia. As a result, (sustainably) grazed grassland sequesters more carbon in the soil than any cropland and even forests. Kris Nichols explained why the act of grazing is so important. When buffalos or cattle graze, 'their tongues will kind of wrap around the plants and pull. That action tugs and rips off some of the root hairs and releases a lot of (sugary) exudates, which immediately triggers some microbial activity.' While they graze, cattle continuously cause such tiny injuries which the grass plants try to heal as fast as they can. To do so they need energy and nutrients and therefore they step up photosynthesis. The plant begins to absorb an increased amount of CO_2 from the air to produce simple sugars which are then delivered to soil microorganisms and fungi in exchange for nutrients. Only grazing triggers this kind of mass photosynthetic action, and that is why (sustainably) grazed grassland[3] sequesters more carbon than cropland and forests. Mowing grass for hay does not have the same effect because there is no tugging, no injury and therefore no need for increased photosynthesis and demand for extra CO_2, says Kris Nichols. 'If we want to address climate change, we need to have grazing animals out there,' she emphasises. To not graze or under-graze grassland is as damaging as overgrazing: very soon, 'bald' spots will appear where the grasses do not regenerate. Instead, invasive species will thrive which produce lots of seed and quickly colonize a whole area. The amount of carbon that can then be sequestered is far less than in sustainably managed and grazed grassland.

A system functions efficiently when the required output is achieved with a minimum of energy. Natural ecosystems rarely work at peak efficiency, they function at the lowest level of efficiency they need to keep going, says Nichols. Only once there

---

3   Grasses also have massive root systems compared to the green blades visible above ground. Their 'root-to-shoot' ratio is different from most other plants (including trees), which have less root mass compared to the leaf mass above ground. The root shoot ratio in grasses varies between 2:1 and 20:1, favouring root mass. As a resul (sustainably) grazed grassland, therefore, most of the stored carbon comes from the grass roots.

is an imbalance, for example because of an injury, energy will be invested to get to maximum efficiency. 'We have the ability to keep challenging our systems to function at peak efficiencies,' Nichols explains. 'And if we keep doing that, we'll get these systems to start storing carbon at a faster rate. If we just let nature do it, that's OK, but it's going to take a long time. If we put energy and resources into it we can change how fast these systems will be able to store carbon and how fast they are going to regenerate.'

The thin dark layer is all the top soil that is left

What Kris Nichols explains here are the same principles that motivated the former French Minister of Agriculture, Stéphane Le Foll, to start the '4 per 1,000' initiative described in the introduction. By now, there are a number of studies, working with slightly different mathematical models, which calculate the carbon sequestration potential of the world's soils.

Pete Smith is a professor at the Institute of Biological and Environmental Sciences at Aberdeen University in Scotland. And he was the lead author of the Climate Change Report at the 4th IPCC (Intergovernmental Panel on Climate Change). In his work, as well as climate change, he focusses on soil biology and carbon sequestration. Pete Smith says how much carbon is sequestered in the soil depends on a number of different factors. 'There is very little point in trying to increase the soil carbon content in well-managed grassland or in a forest because they already have high carbon,' Smith explains. 'The best place to increase the soil carbon content is in crop lands, particularly if they are depleted in carbon because of the constant tillage[4]. And in some severely degraded, mismanaged system like poorly managed grassland, you can improve the carbon stock by improving the management. So really: how much you've lost determines how much you can get back in.'

Kris Nichols predominantly looks at the soils that are typical for the Midwest and the High Plains region of the United States, an area once covered by prairie grasses, and grazed by millions of buffalos. Over millennia, a deep layer of carbon-rich, extremely fertile soil built up underneath the prairie. The process of soil erosion began with the arrival of settlers who plowed up the grassland to grow crops. The tillage released enormous amounts of carbon.

Pete Smith compares the carbon content of soils and the influence of agricultural practices in different climate zones. 'There are two processes controlling the carbon sequestration in the soil, the decomposition of organic matter and productivity,' says Smith. High soil moisture and warm temperatures lead to a lot of plant growth; plants grow fast and produce a considerable amount of biomass. The same conditions, moisture and a warm climate, also accelerate the rate at which biomass decomposes and carbon is lost. That is the reason why soils in the tropics can sequester relatively little carbon: the growth rate of plants is high, as is the decomposition rate of biomass. The result means that it evens out, there is little carbon available that could be stored in the soil long-term. 'You actually find the highest carbon stocks where decomposition is slowed or more or less switched off, in very acidic systems or in boreal systems in the northern hemisphere where it is very cold,' says Smith. 'And there are anaerobic systems where you get a large increase in soil carbon. Peatlands are a fantastic pool of carbon.' Around the globe, there are regions that have an excellent potential for carbon sequestration in the soil, and there are others that have a low potential which cannot be increased. In the tropical belt on either side of the equator, little carbon can be found in the soil. The further one moves away from the equator – to the north or the south – the more carbon can potentially be stored in the soil. Whether the soils in these regions are carbon-rich depends on the degree of available moisture. The decomposition of organic matter is slowed down by dry conditions and accelerated through moisture. Dry conditions also slow down plant growth as less biomass is being produced.

---

4   Tillage turns the soil, deeper layers are moved to the surface, organic matter is exposed to the air and oxidisation sets in, a process that releases CO_2.

Farmers can mitigate climate change in three different ways, says Smith. First of all, there are agricultural practices which keep carbon in the ground instead of releasing it – such as planting cover crops and working with long rotations. You'll read a lot more about cover crops and rotations throughout this book. Different crops need different types of nutrients. If the same crop is grown in the same field year after year, the soil will become depleted of certain nutrients. Farmers can compensate for that somewhat by adding agrochemicals, but with a lack of plant diversity, the diversity below ground will decrease, too. And in depleted soils with fewer, less diverse soil organisms the potential for carbon storage drops. But if the same crop is only planted every five, eight or even ten years, soil quality improves. At the same time, disease and pest pressure go down: weeds and pests may do well one year, but the following year, when another crop is planted in the field, conditions will change too, and neither weed seeds nor the off-springs of a pest will find the same, perfect habitat in which their parents thrived. Their numbers will decline, and they will procreate less. The same goes for weeds: under less favorable conditions, fewer plants will reach maturity and come to seed. As a result, the need for pesticides, herbicides and chemical fertilizer drops.

And that's linked to the second category of measures farmers can take to mitigate climate change: avoiding emission. Producing nitrate fertilizers is very energy-intensive; the manufacture of herbicides and pesticides requires fossil fuel-based energy too. And tractors running on diesel fuel have to be used to distribute all these agrochemicals out in the fields. Every kilogram of fertilizer that is not needed and therefore not produced is an important contribution to climate protection. Less tillage and fewer pesticide and fertilizer applications mean fewer CO_2 emissions.

The third category of climate mitigating practices Smith mentions is carbon sequestration in the soil through permacultures like permanent grassland or forests. But a lot depends on how grass and grazing lands are managed – something we'll get back to later.

Pete Smith assesses the possibilities for carbon sequestration globally and in comparison to the rest of the world. In Europe things look pretty good. More carbon could be sequestered on cropland, but permanent grassland is not routinely overgrazed. However, the management of the large amounts of liquid manure and slurry from concentrated animal feeding operations could be vastly improved as at present too much ends up in ditches, lakes, rivers, and groundwater. 'You could certainly do more in the croplands, things like putting cover crops on, that puts more carbon into the soil during the winter or the autumn,' Smith explains.

'We could use better rotations, we could use deeper rooted crops, and maybe in the future, we could look at the perennialization of our cereals. That would reduce the amount of tillage we need to do, but I think that is something for the future.'

In 2007, Pete Smith and a group of international scientists published a study on 'Greenhouse gas mitigation in agriculture'.[5] 'Many agricultural practices can potentially mitigate greenhouse gas (GHG) emissions, the most prominent of which are improved cropland and grazing land management and restoration of degraded lands and cultivated organic soils'[6], the authors say. But there is no foolproof recipe for how to do that: 'A practice that is highly effective in reducing emissions at one site may be less effective or even counterproductive elsewhere'.[7] When you reach the farm level things get complicated, says Pete Smith when I ask him to give me some examples. 'What we need are tailor made solutions for each and every farm,' he stresses. Soil type, elevation, average rainfall and productivity are just a few of the factors that need to be taken into account. Are there recurring problems with certain pests on the farm, or perennial weeds? Such very specific questions need to be considered too, before a decision can be taken whether carbon sequestration is an option and how it can be achieved. And the authors of the study make another important point: farmers will be much more inclined to use practices known to mitigate climate change if these also increase yield or productivity.

The farmers we visited in the US have started to reinterpret what productivity actually is: it's not about higher *yields* but better *profits*. Soil scientist Kris Nichols agrees. 'You are not looking at maximizing your production, what you are trying to do is utilize the landscape in the most efficient way. Everything will always depend on how much labor you have, how much time you have, what your goals are and what your markets are. You may not do much vegetable production at all if you don't have a market for that or not enough labor,' says Nichols. 'You need to choose the right crop for the right climate, the right soil and for what's needed in the market in your area.'

---

5   Greenhouse gas mitigation in agriculture. Pete Smith et al. Phil. Trans. R. Soc. B (2008) 363,789-813, published online 6 September 2007

6   Smith loc. cit. p. 789

7   Smith loc. cit. p. 798

The Konza Prairie – this is how much of the US looked like for millennia

If these were the guiding principles, US agriculture would change profoundly. For Nichols, it's an absurdity that almost all fruit and vegetables are grown on the east and west coast while the Midwest has become a corn and soy desert. 'Corn is a nutrient and water hog, it is a tropical perennial grass that we annualized to try and grow it as close to the Arctic Circle as possible. How could you possibly think that it's the right thing to do?' remarks Nichols, who comes from Minnesota, a US state in the Midwest, bordering on Canada. Here, too, corn and soy are the predominant crops, and Minnesotans feel the consequences. Nichols tells me about her niece, who belongs to the first generation that statistically has a shorter life expectancy than the generation of her parents. Nichols remembers that in her childhood, every farm had a kitchen garden where the family grew most of their food. Then came a change in the law: students suddenly needed health records and health exams, and that required one member of the family to work off the farm for health care. Usually it was the mother who tried to find work. Teaching and nursing jobs remain very popular in rural areas, not just because teachers and nurses are desperately needed, but also because schools and hospitals automatically sign up the whole family for health insurance. Suddenly, there was no time left to tend the garden. Her niece's generation doesn't know how to grow vegetables, says Nichols. Today, the vegetables on the plates of Midwesterners come from a can, from a bag in the freezer, or they were grown in California and have been transported over thousands of miles. 'We're now facing a health crisis where people are obese and malnourished at the same time,' says Nichols.

The nutrient density in fruit and vegetables continues to decline, and we need to eat a lot more to supply our bodies with the same amount of minerals, vitamins and trace elements. Transport plays an important role; the nutrient density begins to decline as soon as fruit and vegetables have been harvested. But soil quality matters too: antioxidants like flavanol, beta-carotene and lycopene as well as the vitamins A, C and E are stress molecules, says Nichols. They are part of the plant's defence system against pests. 'Some of the antioxidants and amino acids that the plants have are actually produced by soil microbes. The only way plants can get them is from the soil,' Nichols explains. In degraded soils, with few soil organisms, not enough nutrients are available and the plants cannot produce enough antioxidants. For fruit and vegetables on supermarket shelves to look perfect nevertheless, farmers have to use large amounts of pesticides and fungicides[8]. 'We have to grow food again,' says Nichols. 'We have forgotten that farmers grow food, not feed and commodities.'

To Nichols, nutrient density and soil quality are much more important than yield. And in her opinion, the nutrient density of fruit and vegetables might be a better indicator of soil health than the amount of carbon stored. In a soil analysis, it is difficult to tell labile and stable carbon apart. And the carbon content in soil does not increase in a linear fashion. Pete Smith named temperature and rainfall as two important factors that impact the potential for carbon sequestration, which means that the soil carbon content can vary from year to year even if farmers do everything correctly and employ best agricultural practices.

In principle, Nichols agrees with former French Minister of Agriculture Stéphane Le Foll and his '4 per 1,000' initiative. But she doubts whether it will be possible to define a global standard. The more degraded the soil is, the easier it is for a farmer to change to better practices and increase the carbon content. If carbon sequestration were to be financially rewarded, the farmers who have been working for many years to improve their soils should not be disadvantaged because the carbon content has plateaued, says Nichols. And there is another aspect: 'The

---

8   A 2015 Greenpeace study showed that in a conventional apple plantation, there are on average 32 pesticide applications in a single growing season. https://storage.googleapis.com/planet4-international-stateless/2015/10/1a0d04c1-europes-pesticide-addiction.pdf

average farmer understands supply and demand, selling and marketing things. Farmers feel good about growing things that have value. They can feel good about being paid for conserving the environment – but being paid for carbon? Farmers are "action people",' says Nichols, and she can't quite see how carbon sequestration would fit into how they see themselves.

I asked Kris Nichols what needs to happen for farmers to do more to improve their soils and thereby sequester more carbon. She doesn't hesitate: top priority is a functioning infrastructure system which allows farmers to diversify and sell a large number of different products rather than trying to increase the yield of a single produce, like soy or corn, as is the case now. For farmers to switch from conventional agriculture to a regenerative system, some financial support would be needed. Changing how they work in such a fundamental way holds enormous risks in particular in the first few years. Without any kind of financial protection or insurance, most farmers are not willing to take such a step. Government programs should incentivize regenerative agriculture and the production of food that deserves the name, says Nichols: 'Our farmers don't feel as connected any more to the soil or the land because they forget they are growing food and are treated as if they were just growing commodities.'

The connection between climate change, agriculture, and the quality of food also must be made a priority for research. For Nichols, more emphasis needs to be put on field research and less on what is statistically significant and what is not. And she would like to see teachers and educators everywhere making the connection between climate change and agriculture, and what it means for the availability and quality of our food. All of the above has to happen fast, of that she is sure too: 'Modern conventional agriculture is an incredibly fragile house of cards, and it is in no way sustainable. I am not sure if in some way indirectly the climate crisis is going to cause this house of cards to collapse far faster than anyone thought possible.'

Farming with nature and building communities

Agrochemical companies say that it is up to us to decide what all our lives will look like. Advertisements for fertilizers, pesticides and other agrochemicals lay out clear choices: we can opt for a golden future in which content farmers produce an abundance of food for discerning customers. Or we can reject genetically modified seeds, fungicides, herbicides, and other agrochemicals, and we will soon live in a world full of hunger and poverty because we cannot feed the growing world population. Without products from companies like Bayer-Monsanto, Syngenta-ChemChina, Dow, and DuPont we have no future, says the industry. If we tried to farm without them, they say, we would never achieve the yields needed to satisfy the rising demand.

Amish country

Advocates of organic and regenerative agriculture beg to differ. They argue that in the long run, it will be industrial agriculture, dependent on agrochemical inputs, which will be unable to feed a growing population. Very soon, nothing but herbicide-resistant 'superweeds' will be able to grow on degraded soils, and no amount of chemical fertilizer will change that. Both sides argue that their views are based on sound science. Results from lab trials will do little to decide the controversy one way or the other. What's needed is strictly controlled field trials that are conducted over a long period of time. The longest such trial is being run at the Rodale Institute for Organic Research in Kutztown, Pennsylvania.

As the crow flies, Kutztown is just 100km (65 miles) to the north of Philadelphia. The scenic route through the rolling hills of Lancaster County is a bit longer, leading along narrow country lanes flanked by hedges. The detour is well worth it. There are few places I can think of where the contrast between modern life and a seemingly bygone era is as stark and direct as it is here. Some 300 years ago, the members of a Christian sect, the Anabaptists, emigrated to America for religious reasons and settled in Pennsylvania. To this day, the Amish or Pennsylvania Dutch have kept their style of life. Men wear long beards and keep their heads covered with a (straw) hat. Their means of transport is horse and buggy. They can be seen everywhere, moving along at a fast clip with a line of cars behind them, parked outside a store or at the drive-through of the local bank. The women wear bonnets and long skirts with an apron, and can be seen on kick scooters – bicycles are not allowed as they could be a temptation to go somewhere just for fun, and not because it is strictly necessary. Most Amish are dairy farmers, they produce fresh milk, but also cheese, yoghurt, and ice cream. On their fields, they use neither chemical fertilizers nor herbicides. Cow manure is spread as fertilizer, while the plow is drawn by one or two horses. An Amish farm is easy to recognize – not just because of the traditional red barns but because of the missing power lines; the farms are not connected to the grid.

And as if the clash between this bygone world and the modern one was not dramatic enough, the border of Lancaster County to the west is marked by the Susquehanna River. Even many Americans will not necessarily be familiar with the river's name, but they will probably have heard of one of the islands in it: Three Mile Island. It's the site of the atomic power station where a serious nuclear accident occurred in 1979.

It's not just the Amish living in such proximity to the Three Mile Island nuclear reactor which can cause spells of severe dissonance in the traveler. We decided to take the short route to get from Philadelphia to Kutztown and the Rodale Institute, straight up Interstate 476 and past Allentown. I remembered the name from a 1980s Billy Joel song. 'Out in Bethlehem they're killing time' was one of the lines that didn't make sense to me then. Suddenly the Bethlehem exit comes into view, and a little while later we are standing opposite the gargantuan structure that used to be the Bethlehem Steel Works. Until its closure in 1995, it was the largest steel works in the US, having been in operation for 140 years. Today, the huge smelters, furnaces, chutes, and pipes are rusting. The elevated rail track on which freight cars once delivered iron ore and coal from Pennsylvania mines directly to the furnaces has been turned into a walkway from which visitors can peer into the old factory buildings. Young trees, shrubs and grass grow where more than 30,000 people once used to work. It is a rainy day, and apart from us there are hardly any visitors. It's eerily silent. For a moment, it feels as if I've suddenly landed in some post-apocalyptic future in which nature is reconquering a world without humans.

Bethlehem Steel near Allentown, once the largest steel works in the US

It is easy to read too much into such an image, to overinterpret its symbolism – there is no connection between the decline of Bethlehem Steel and industrial agriculture. But as it's hard to enjoy Hawai'i's beaches[1] once you know about what the agrochemical industry does on the other side of the islands, in Pennsylvania it is impossible to ignore the contradictions between Amish, nuclear power plants and Allentown.

The Rodale Institute is situated a few miles outside of Kutzdown. A tree-lined country road leads up to the beautiful old farm buildings which today house offices and classrooms. The Rodale Institute not only conducts organic research and field trials, it also offers consultancy work, internships for American and

---

1   See chapter 4

Farming with Benefits

Bethlem Steel – or what's left

international students and various courses in organic agricultural practices. To promote organic and regenerative agriculture amongst consumers, Rodale has launched a new certification scheme. The Regenerative Organic Certification (ROC) sets standards that go well beyond what is needed for USDA organic certification. ROC combines organic agriculture with very high animal welfare and environmental standards as well as fair wages for agricultural workers. Rodale cooperates with a small group of farmers, the USDA, animal welfare and environmental organizations and a number of companies that are already promoting high environmental and social standards, like the outdoor clothing company Patagonia or the soap manufacturer Dr. Bronner. Products qualifying for the ROC label will be considerably more expensive. 'They will only appeal to very discerning customers,' says Diana Martin, who coordinates the project at Rodale. In the long run, she hopes, the label will help to set higher standards internationally.

While new projects are being developed, old ones continue. We are on our way to see the field trials that were started at Rodale in 1981. A fence and a hedge separate the 12-acre test site from the rest of the farm. The trial compares different organic systems (one working with manure, the other with legumes only) to conventional agriculture. On the conventionally managed field strips, fertilizers and pesticides are applied in accordance with the state agronomists' recommendations for this part of Pennsylvania. In 2008, a no-till trial was added to the organic management systems, and only half of the strips that are managed with manure or legumes respectively is still being tilled as before, while the other half is not tilled. And on some of the conventionally managed test fields, genetically modified seed varieties are planted; here, too, half of the field strips is no-till. In both systems the rotation is typical for the region. On the conventional fields, corn and soy alternate, and since 2004 wheat has been included in the rotation. In the organic system, corn, soy, and wheat are followed by a year which is designed purely for the benefit of the soil: in one trial, red clover and alfalfa are grown and harvested as hay, in the second trial legumes are planted as cover crops, and the third trial combines cover crops and manure.

There are 72 test plots. Such a large number is needed to accommodate the differences between the conventional and organic systems, the different management practices within each system, and the rotations.

In 2015, a group of scientists analyzed and compared the existing long-term organic field trials in the US[2]. For a long time, the assumption was that yields in organic agriculture would lag considerably behind those achievable in conventional systems. The scientists concluded that this only holds true during the first few years. Once the transition period is over, the yields in both systems are similar. In drought years, organic agriculture is the clear winner: in the Rodale trial, yields in the organic systems were up to 40 percent higher than in the conventional ones[3]. And contrary to expectations, there was not a single year in which genetically modified corn did better than conventional non-GM varieties. Water analysis showed more nitrate run-off from conventionally farmed soils. In the organic system, soil analysis showed marked differences in soil quality: using manure as fertilizer resulted in more carbon sequestration than planting legumes. In fields treated with manure, 981kg (2162 pounds) of carbon per hectare was stored in a year; in the legumes-only system it was 574kg (1,265 pounds). Conventional agriculture needs 45 percent more energy. In the trials, this was mostly due to the use of chemical fertilizers and their energy-intensive production. Overall, the greenhouse gas emissions in conventional agriculture are 40 percent higher than in organic agriculture.

---

2   https://www.researchgate.net/publication/281511510_A_Review_of_Long-Term_Organic_Comparison_Trials_in_the_US

3   https://rodaleinstitute.org/science/farming-systems-trial/

The Rodale long-term trial uses small plots managed in standardized conditions. An experienced organic farmer will be flexible and able to react in a timely and appropriate way to factors such as weather conditions, temperatures or market conditions and thereby can further improve yields and profitability.

On the way back from the test fields to the main building, we spot several of the famous 'roller-crimpers', a tool developed by Rodale CEO Jeff Moyer. It's specifically designed for use in organic agriculture and any farm working with cover crops. To ensure good optimal growth and yields, the competition for light and space in a field has to be removed before crops like corn, soy or wheat are sown. Cover crops sown in fall had all winter to grow. By spring, they would be tall and deeply rooted. In conventional agriculture, herbicides are used to 'terminate' them. Farmers with a cattle herd can graze them, some cover crops can also be mowed and used as fodder. Organic farmers who don't keep animals, however, have a problem – one that Moyer tried to solve with the roller crimper. It consists of one or more heavy, cylindrical rollers which are drawn across a field by a tractor. Each roller is fitted with short, sharp metal edges arranged at an angle. In one go, the roller crimper flattens (rolls) the plants while the metal edges simultaneously break and squash (crimp) them. If conditions are right, the roller-crimper can be run in front of the tractor while a seed drill for direct seeding runs behind[4]. The method saves not just time and fuel but also protects the soil, for fewer pass with a tractor mean less compacting of the soil and less harm to the soil life. We learn what it's like to use a roller crimper two days later, when we visit Klaas Martens who farms in upstate New York.

For Americans this will immediately conjure up images of lakes and a beautiful, rural landscape. Outside of the US, most people tend to think of the city first rather than the state which, in the context of this journey, makes for a rather unusual point of reference: New York City is situated in the very south of the state, at about the same latitude as Allentown. The state of New York reaches far west and north – beyond the border lies Canada. In the middle of June, the landscape in upstate New York reminds us of the area around Lake Constance in Europe, with rolling hills, light forests, and narrow, meandering roads. What's American is the farmhouses and barns, wood-paneled and painted red with one

---

4 For direct seeding, the field will not be tilled; instead, the seed is put directly into the soil. After a pass with the roller-crimper, the soil is covered with a thick layer of plant residue, and special equipment for seeding is needed: it cuts through the plant material to make a furrow into which the seeds are deposited in the specified intervals. There will be a lot more about this technique in later chapters.

or two tall, slender grain silos in the yard. And then there are the lakes, long and narrow, like valleys filled with water. Lake Seneca is 38 miles (61km) long. Just a few miles to the west another lake runs parallel and on its northern edge sits the small town of Penn Yan.

When we pull into the yard, Klaas Martens and his son are standing next to one of the huge steel grain bins, checking the quality of the freshly harvested malting barley. It's one of about 30 crops the Martens family grow on their certified organic 1,600 acre farm.

For Klaas Martens every field tells a story

The first time I heard of Klaas Martens was in January of 2014, when he was interviewed by Melinda Hemmelgarn, a journalist and registered dietitian based in Columbia, Missouri. Once a week, she hosts the 'Food Sleuth' podcast[5] in which she 'connects the dots between food, health and agriculture' and encourages people to 'think beyond their plate'.

Martens told Hemmelgarn about returning to the farm after university. At the time, the way his father farmed seemed backward and out of date to him. With hybrid seeds, chemical fertilizers and the new generation of herbicides that had just come on the market he intended to bring about a 'green revolution' on

---

5 https://beta.prx.org/stories/109184

the farm overnight, Martens recalled. And then he tells Hemmelgarn how one morning, he had been spraying a mix of 2,4-D[6] and other herbicides. He had come down from the tractor to fold up the sprayer and move to another field but suddenly found he couldn't move this right arm. Martens said he has no proof that the herbicides he sprayed that day caused the paralysis, but common sense suggested that this had been the case. 'You reek of poison,' his wife had told him when he came through the door. He slowly regained movement in his arm, but his hand stayed paralyzed for the rest of the summer. He was unable to continue spraying, and for ethical reasons he felt he couldn't hire anyone to do so for him. He started thinking about how often his work overalls, reeking of pesticides, had gone into the washing machine together with children's clothing. 'Am I really prepared to poison the kids? he'd wondered. Soon afterwards he and his wife, Mary-Howell, took the decision to convert the whole farm to organic. He didn't come to organic farming as a complete novice. In a farming magazine he'd read about a processor who was desperately looking for growers who could supply organic wheat. Martens became curious and started to do some research. At nearby Cornell University, he found books published in Europe in the 1920s and 30s that talked about weed suppression through long rotations. Martens drew the reverse conclusion: if American farmers, year on year, had grown corn, soy and wheat as monocultures, they had also created ideal conditions for weeds. He decided that he wanted to try out alternative methods and see for himself whether they worked. To Martens, research and farm trials belong together; how much knowledge gained in a lab is worth can only be assessed in the field.

June is a very busy month on the farm, but Klaas Martens still makes time for our visit, and within minutes we're deep in conversation. 'We've only been measuring yield; we didn't look at the effects on the soil, on nutritional value on crops like wheat, on the ecosystem. Scientists saw what they wanted to see and told us to ignore anything that didn't fit,' says Martens. He is very taken by Rudolf Steiner and the principles of biodynamic farming[7]. 'He was such a keen observer and looked at systems rather than one individual aspect.'

---

6   2,4-D was the active ingredient in 'Agent Orange', a defoliant the US used during the Vietnam War.
7   Biodynamic agriculture is based on the concepts developed by the Austrian philosopher, economist and social reformer Rudolf Steiner. Key are homeopathic preparations to enhance soil health. Animals, in particular cows, are considered essential for biodynamic farming, and the phases of the moon are relevant for deciding when fieldwork should be done. The standards for biodynamic agriculture are based on those for organic farming, but are much stricter. In Europe certified biodynamic products carry the Demeter label.

Klaas Martens beams when he starts talking about agriculture, science and philosophy; the farm walk has turned into a lecture, and I'm the student. He peppers me with questions to find out what I notice about plants and soil and what conclusions I draw. To him, walking the field is a little adventure. 'It has to be fun,' he says as he discovers something else that has him raise yet another topic and point out more connections.

We've reached the first of the fields he wants to show us. Over winter Martens grew rye as a cover crop, and then he sowed soy. The right half of the field looks much better than the left; the young bean plants are developing well and seem a lot stronger than on the other side of the field. On one side, he mowed the rye as a fodder crop, says Martens. On the other side, he used a roller-crimper, the farm tool we'd seen at the Rodale Institute. It's left a thick layer of organic residue on the field, and the soybean seedlings have a hard time pushing through. Economically, too, it makes sense to harvest cover crops as fodder. The yield from 10 acres was just under 1,000kg (2,000 pounds), enough for two cows. At present, the fodder isn't needed on the farm, but there are a number of dairy farmers nearby who are happy to buy it.

On to the next bean field. Here, Martens sowed mustard over winter which 'cleanses' the soil and helps protect the next crop, soybeans, from root rot. When growing cereals like wheat Martens underseeds with legumes like clover which can fix nitrogen in the soil and make it available to neighboring plants. Underseeding is one way in which farmers can make sure that their cash crop gets enough nutrients without adding chemical fertilizer.

I ask Klaas Martens what his take is on Kernza, the perennial wheat scientists at the Land Institute in Kansas have developed. As it turns out, Martens is one of the farmers taking part in a trial to grow Kernza commercially, and he is working closely with the scientists conducting the research and collecting the data from growers in different parts of the US.

All cereals we grow today originate from wild grasses. Since the first farmers started cultivating land, seeds have been selected for kernel size, yield, and other advantageous traits. Over many centuries, distinct grain varieties like wheat were developed. Compared to wild grasses, they became highly productive, but they ceased to be perennial. To grow cereals, farmers plow to prepare a seedbed, and over time this will degrade the soil. The ideal solution to the problem would be a perennial wheat. In the early 1980s, scientists at the Rodale Institute started to identify prairie grasses which potentially could be advanced through breeding. The first trials started in 2006.

The Martens' seed and grain business next to the railway line in Penn Yan

When I visited the Land Institute in 2013, research director Tim Crews showed me images of prairie grass roots and the roots of wheat plants. The contrast is stark: wheat plants only have one short season to develop roots. They are short and shallow because the plant invests most of its energy into producing a large number of ears with big kernels. The roots of perennial prairie grasses have years in which to grow and can reach several meters deep into the soil. As a result, they are much more drought-resistant. The roots have access to moisture in deep soil layers that the short wheat roots cannot reach. And because prairies are plowed (unless they are converted into cropland and stop being prairies), a layer of humus develops in which carbon is stored long-term.

A perennial wheat variety would have a lot of advantages for soil health and the environment, as long as the yields were high enough to make it economically viable. The scientists at the Land Institute started to select the prairie grasses for a number of strict criteria: high yield, large kernels ripening at the same time, low shatterability (the grains stay attached to the ear and don't fall out easily) and homogenous plant height. The best-performing plants were crossed with annual wheat and the result is Kernza, a perennial wheat. In 2013, Tim Crews was optimistic. Kernza grains only are about a third of the size of ordinary wheat grains, and it would take an enormous amount of land to grow commercially viable volumes of Kernza. But a flour mill in Kansas had already milled Kernza into flour which had been tested by several bakeries. Like rye, Kernza contains little gluten, but the bakers found that combining equal parts of Kernza and traditional wheat flour is ideal for baking bread which, according to those who had a chance to try it, has a very pleasant, slightly nutty taste. Brewers, too, have tried Kernza to brew craft beers. In 2017 the US multinational General Mills announced a cooperation scheme with the Land Institute[8] to up production and work with processors in order to establish Kernza in the mainstream market. General Mills also financially supports research at the University of Minnesota; extension agents work with farmers across the northern US, where growing conditions for Kernza are better than in Kansas.

At Klaas Martens' farm in northern New York, Kernza should be doing well. He shows us the small field in which he has been growing Kernza for the past three years. It takes us a moment to understand that this isn't just a patch of ordinary grassland. The seed grains are tiny, and still have to be de-husked. Once that's done, there's practically nothing left, says Martens. To get any kind of yield requires a lot of chemical fertilizer and heavy use of pesticides. With such inputs, a farmer can expect about 90kg (200 pounds) of Kernza per acre, whereas wheat would yield ten times that amount. But Martens farms organically and does not use agrochemicals, which explains why the results on his test plot are so disappointing. 'Up to now, Kernza is nothing but grass seed and suitable only as fodder,' says Martens. He holds little hope that plant breeders will be able to significantly improve Kernza to the point that growing it will make commercial sense. And Kernza would not be the first perennial grain. Perennial rye has been available for a while, says Martens, and growing it is a commercially viable alternative. The problem remains with the marketing – there is far less demand for rye than there is for wheat.

At General Mills, the initial big plans for Kernza have been somewhat scaled down[9]. Cascadian Farms is one of General Mills' organic food brands. In 2018, they had started to develop a Kernza-based cereal, but the yields that year were so bad that only 6,000 packs of cereal could be produced. In April of 2019, General Mills announced that a limited edition of Kernza cereal would be distributed among customers making a donation of $25 or more to the Land Institute in Kansas. Given the low volume of Kernza available, plans for a nationwide rollout

---

8 https://apnews.com/b4aa1f1eabcb46b5b5cefe83e31092c7
9 http://www.startribune.com/kernza-crop-failure-sends-general-mills-unit-to-remake-plans-for-new-cereal/508349702/

of Kernza products could not be realized. For that to happen, the size of Kernza grains and the yields would have to be considerably improved.

Klaas Martens has chosen a different route to work with the environment and stay profitable. We drive to a field Martens calls his field lab. It is surrounded by woods on three sides, and the previous owner simply abandoned it after managing it conventionally over a very long period. The soil is degraded, and only weeds seem to thrive. Crops were prone to disease and acted like a magnet for every conceivable pest. 'Farmers ask the wrong questions,' says Martens. 'When their agronomist sees weeds, he says: kill them. The right question is: why are they here? What can they teach me? It is the attitude: Do we see something that needs to be killed or do I see an imbalance? Martens has walked a few meters into the field while naming the plants he spots. 'Every field tells a story,' he says and starts translating the narrative for me: chamomile indicates a lack of calcium. But a soil analysis is needed to decide whether the soil really lacks calcium itself or whether it lacks the microbes that 'prepare' calcium for the plants. And there is no need to eradicate chamomile, says Martens; it will wilt and have no negative impact on the growth of other plants. A number of different plant species, all with yellow flowers, indicate the presence of sulfur. One of them is ragweed; it can grow quite tall, but nowhere near as tall as the glyphosate-resistant plants ubiquitous in many fields in the Midwest. Then there is sparrow vetch – not a problem, says Martens, as it will never catch up with a thriving soybean plant. Summer vetch indicates that there is not a lot of nitrogen in the soil, which means soybeans will do well here. Martens names more 'indicator' plants and what they tell him about the field. It's been four years since he started observing this particular patch of land. In the first year, the weeds ran riot. In the second year there were fewer weeds because the soil environment had changed: the weed seeds had to contend with an environment that was a lot less favorable than the one in which the parent plants had done so well. Off-spring of the weeds that were happy there the year before didn't find the same environment. In the third and fourth year, perennial plants like golden rod and other, woodier species began to move in. In fall he planted rye and peas as cover crops which started to regrow in spring. The Martens' have been harvesting the pea shoots and added them to salads. There are so many peas that harvesting the shoots and selling them in a farmer's market would be a viable option, but it's the type of crop that doesn't quite fit into the farm's system. The biggest problem for crops of any kind in this particular field is four-legged and comes from the surrounding forest: deer cause thousands of dollars of damage every year. They cannot be hunted until fall, and in Martens' opinion, the state doesn't do enough to control them.

The conclusions Martens draws from the observations in his 'field lab' help him to develop new strategies; the effects of the changing climate have been obvious for a number of years.

Checking grain quality

The barley harvest has moved forward by three weeks, which also has advantages: the growing season has become longer, which makes it possible to grow a second crop, such as pinto beans, once the barley has been harvested. Unfortunately, the changes have also created favorable conditions for pests and diseases that farmers in the region never had to deal with before. Martens stays with a systemic approach: 'For every intractable pest on the farm that tempts us to use pesticides there is a species that, if introduced at the right time, will eliminate the pest. Pests are nature's attempt to balance an imbalance that we have created.'

In regard to climate change, Martens sees rain events as the biggest problems – torrential rains that happen more and more frequently, dumping enormous quantities of water in an extremely short period. 'In 2000, we had 8 inches (200mm) of rain in a single day and thought this was a once in a lifetime event. In the last few years we've had 5 inches of rain within an hour, not once, but several times,' he recalls. Worst of all are torrential rains in February or March, when the soil is still frozen and unable to absorb any water. In such situations, cover crops are the only remedy: the living roots hold the soil. Martens has seen clover with bare roots that seemed to sit on a tiny island – the water had washed away the soil surrounding it[10].

The amount of stable carbon in the soil is directly proportional to the amount of water the soil can hold, says Martens. In New York State, the average amount of soil organic matter varies between 1 and 1.5 percent. On Klaas Martens' farm, the average is three percent, and on fields that have not been tilled for up to 20 years, the soil organic matter is as high as four percent. Martens believes that there is a carbon equilibrium in a long-term farming system: the carbon level will go up or down and essentially fluctuate around a certain level. He does not think it is possible to continuously increase the amount of carbon, but what can be increased is the humus layer. 'We could increase the amount of topsoil by breaking into the subsoil. Subsoil tillage[11] makes deeper soil layers accessible to plants if you immediately plant something that grows deep roots which will then extend into the newly tilled subsoil area,' Martens explains. He would like to do a trial with Cornell University using a canola variety that can grow 60 cm long roots in just 45 days. 'We need to quantify what we think works,' he notes. To supply the canola plants with enough nitrogen Martens would seed it together with clover, a combination that would also make an excellent fodder crop. 'Tillage is like drawing money out of the bank; it can be good if you know what you want

---

10 Torrential rains are the new normal for farmers in North America, as are periods of extreme drought. In the first half of March 2019, extreme rainfall events in several Midwestern states caused widespread flooding and millions of dollars in damages. Among the hardest hit regions were Nebraska and western Iowa. Roads and bridges were destroyed, and on some fields, all topsoil was washed away. Many levees broke, and rivers carved their paths through fields and open countryside before eventually returning to the original river beds. Once the waters receded, what was left were tree trunks, boulders and waste. Some farmers found their fields covered by a layer of sand and silt up to a meter thick. It will take years until anything can be grown again on such land. As commodity prices had been depressed for a while, farmers had stored record amounts of corn, wheat and soy on farm in bins and grain bags. The power of the water was such that silos were ripped from their foundations and carried away by the floods. Other grain bins were flooded, spoiling the stored grains. Thousands of cattle drowned, among them many new-born calves. The unseasonal, heavy rain lasted for several weeks. As a result, about a third of all acreage in the Midwest could not be planted at all in 2019. Where farmers did manage to get seed into the ground, yields were expected to be well below average.

11 This method differs from deep plowing, where the plow shares not only reach deep but also *turn* the soil, so that deep layers are exposed to air and weather while the top layer is worked deep into the ground. Subsoil tillage breaks up compacted subsoil layers but does not turn the soil.

to invest in,' says Martens. In his opinion, no-till has become something of a dogma, to him it's important to know when to plow and when not to. Martens plows occasionally, usually before planting corn or soy. In his experience, tillage helps to warm the soil – something that can be very important in a region with long, harsh winters and a short growing season.

If there is a large amount of organic matter in the soil it gets shifted to a deeper level. Klaas Martens' son Peter thinks that this is beneficial for root development, a theory he would like to test in a field trial and in cooperation with scientists from Cornell University. Whether or not plowing is beneficial also depends on the soil type, says Martens. The soil in the Penn Yan area contains up to 50 percent of silt. Silt particles are tiny and dissolve in water, unlike clay or sand. When it rains, the silt infiltrates the soil with the water and gets carried to deeper soil levels, where it deposits and forms a hard pan. This dense silt layer not only blocks off oxygen and prevents growth, it is so hard that neither roots nor earthworms can break through. It's this silt barrier that Martens breaks up through plowing every few years.

In our conversation, Martens puts details into a larger context; he develops hypotheses only to immediately test them against observations. For him, nothing, no event, no action, no observation, stands on its own, everything is connected, part of a system, or a network. System, network and community are synonyms for the same principle. Farming does not happen in isolation; for Martens, the community in which he lives, the neighboring farm families, people living in Penn Yan, are just as important. On the farm, he promotes soil health by looking after the community of soil organisms that sustain the growth of his crops. Off the farm he cooperates with his neighbors, helps and supports them. For him, everything hinges on community and what he calls 'community knowledge': in a group individual knowledge starts to add up and develop into 'community knowledge', a valuable resource that, in turn, defines the character of a community. Klaas Martens' philosophical musings are part of the discussion at lunch, which we share at the huge wooden table in the farmhouse kitchen. One very practical way to build community and community knowledge is the 'agricultural circle' Martens organizes during the winter months. 'We sit in a circle because we are all equal and then we talk about issues and try to come up with solutions,' he says. 'Sometimes the decision is taken that more information and training is needed – for example how to read and interpret a soil test correctly – but we don't give recipes, we don't have them, it is all about observation and sharing.'

In fall of 2017, the Martens experienced what being part of a functioning community truly means. At the time, the big barn just across the yard still housed hogs and dairy cows. On the evening of the 23rd September when she was just about to go to bed, recounts Klaas Martens' wife, Mary-Howell, she looked out of the window and saw that the barn was ablaze. The animals were housed in one part, while the other side was used for the storage of hay bales and machinery. To this day, the Martens don't know how the fire started. Within minutes, the first neighbors showed up, then the firemen started to arrive, not just those living in Penn Yan but 'every fireman in the whole county'. It took them all night to contain the blaze and keep it from spreading to the other buildings. The cows got out, but most of the pigs didn't and died in the flames. Mary-Howell has tears in her eyes as she recalls how she and other women tried to help the pigs which had been rescued but had suffered third degree burns. Some were so badly injured that they had to be put down. 'It just was a horrible sight,' she says. It took several days for the insurers to assess the damage, and until they had finished, the site couldn't be cleared. But on the 1st October, just a week after the fire, the remains of the old barn had gone. Early the next morning the neighbors showed up with tools and building material and started to raise a new barn. By the time the first severe frost set in, the new barn was ready, and the cows could be housed. As Mary-Howell tells the story, everyone at the table gets rather emotional, once more reliving that fateful night. We, too, are quite shaken. But for Mary-Howell, it's not just a tale of misery: 'When people said they were sorry for our loss I replied that it was terrible, but that we really know that we have a community, that we live among people we can trust.'

The Martens not only farm, they also run Lakeview Organic Grain, a seed business and feed mill. Selling and processing grains perfectly compliments the farming enterprise, but when they took the decision to buy the mill in 2001, other considerations played a much larger role. In this part of New York State, there are still a lot of small dairy farms, and the demand for cattle feed is high. Commercial feed companies prefer to work with big farming operations and sell large volumes. From the late 1990s onward, the Martens started getting more and more calls from farmers with just a few animals. They were hoping to buy small amounts of feed, and Martens supplied them the best he could, although it was a 'seat of your pants' operation, says Mary-Howell. 'When the local seed trader went bust and we decided to buy the mill, we saw it as a community resource that benefits us all,' she explains when we first meet her in her office at the mill in Penn Yan. The beautiful old building that houses the mill could do with a fresh coat of paint, but that's just looks. The mill itself is in perfect condition and pretty much working non-stop. Initially, the Martens milled about 200 tons of grain per month, whereas today, on average, it's 1,000 tons. Lakeview Organic

Amish customer at Lakeview Organic

Grain sells feed mixes for dairy cows, hogs, chickens, sheep and goats, in bulk or packed in bags. The shelves in the combined shop and office space would not look out of place in a heath food shop: farmers can pick up vitamins, minerals and any other supplements cows, sheep and chickens need to stay healthy. With feed, farmers have a choice between the 'house mix' and bespoke feed and mineral blends. Lakeview Organic has a small fleet of delivery trucks but also works with shippers who truck for them. Most customers are in New York State, but farmers in Massachusetts, Vermont and even Pennsylvania place orders too. And requests keep coming in from further afield, like northern Maine. 'We'd love to help the farmers there but it doesn't make sense; because of the haulage charges there is no way the mill could supply at a reasonable rate,' says Mary-Howell. Lakeview Organic works closely with 50 organic growers in New York, Pennsylvania and Ohio. 'We need to be confident of their integrity, Mary-Howell adds. Farmers often don't have storage capacity or drying and cleaning facilities on the farm. Usually, Martens buys the grain as it comes out of the field and pays immediately (in full or an agreed amount) on collection. For growers, it's rare to be able to sell under such favorable terms. Very often the buyers contractually oblige the farmer to store the grain on a farm for up to a year, and it can take months for payments to be made. Such terms force farmers to invest in grain bins, and when payments are delayed, there is often no choice but to take out an expensive short-term loan.[12]

Apart from feed, Lakeview Organic offers seed, too: to buy organic seed is difficult, and to obtain small quantities can prove nearly impossible as demand far outstrips supply. 'I get a lot of calls from farmers who want to convert to organic,' says Mary-Howell, who spends much time advising and supporting them. Additional information and help are available on the website. 'The mill is used as a tool to create community and help people farm organically,' says Mary-Howell.

Later, during lunch in the farmhouse kitchen, Klaas Martens provides more context. In 2014, only 8 percent of agricultural land in the county was managed organically, now it is 75 percent. 'Neighbors see how we work, they get interested and want to do the same', says Martens. To help farmers convert to organic is extremely important for Klaas and Mary-Howell as they want to do right by the community, the human one above ground as well the community of soil organisms below. And in this part of New York State, community still has yet another meaning: in the last 30 years more than 700 Mennonite and Amish families have left Pennsylvania and settled in the area around Penn Yan. Farms are not automatically handed to the oldest son or split up. Instead, the families try to acquire land so that all children who want to farm can do so; the alternative is an apprenticeship to learn a craft. In the 1990s, Amish and Mennonites in Pennsylvania ran into problems trying to buy farmland; there was little for sale and prices increased dramatically. The two communities therefore decided to join forces and look for a region in which to settle and acquire land. Around Penn Yan, they found what they had been looking for. Today, men with long beards and women wearing bonnets and long dresses are a common sight in the town and at Lakeview Organics. Amish and Mennonites have opened workshops and small businesses, they have founded schools, they volunteer as firemen, they bake bread, make cheese and jam and sell it at local farmers' markets together with seasonal fresh produce from their gardens. In New York State, too, rural flight was a longstanding problem, with many rural communities becoming virtual ghost towns. But in Penn Yan and other small towns in the region, new shops have opened, and there are banks, schools and service providers. For religious reasons, the Amish are not allowed to operate motor-driven machines or equipment. That has opened up new business opportunities for non-Amish who offer services, including transport, Internet access or use of machinery.

---

12 More about marketing and farm infrastructure in chapter 7.

Amish and Mennonites are excellent farmers and business people; a 100 acre (40 hectares) farm is enough for a family to make a living. And while farmers in the US on average are between 58 and 60 years old, the average age in Penn Yan and Yates County is 35. The yellow road sign with the stylized horse and buggy that alerts motorists to Amish carts and carriages on the road is also an indication that the next township probably is a flourishing and prosperous community.

Klaas and Mary-Howell's son Peter and daughter-in-law Hanna will shape the future of the farm. For a long time, the Martens have grown a large number of different grains, among them varieties ideal for bread baking. They are delivered to a grain mill in nearby Ithaca and ground into flour. Most of it is snapped up by Wegemans, a supermarket chain with 90 outlets across six states and a large organic product range. Could the Martens imagine milling their own bread flour? Lake View Organics doesn't have the right milling equipment, says Mary-Howell and the health and safety standards for producing food grade flour are extremely high. Through their cooperation with the mill in Ithaca, Klaas Martens knows that the millers there have undergone specialized training and by now have plenty of experience in handling organic grains. In addition, the mill has set up a 'bread lab' which tests different grain varieties and mixes for their baking properties and taste. Son Peter has done an apprenticeship to become an artisan baker. He knows how to work with traditional sour dough, long proving times and grains like emmer, khorasan and spelt. In future, he would like to clean, pack and store bread grains on the farm so that they could be sold directly to small bakeries who mill their own flour as and when they need it. And the way the small towns in Yates County are thriving and developing it wouldn't be surprising if more artisan bakers were to set up shop in the region.

Grains are our oldest and most important food crops. Often, the course of history depends on whether they flourish or fail. A little over a hundred years ago it was wheat that changed the history of the United States.

Farming with Benefits

The 'Dust Bowl' – a history lesson not learnt

When industrial agriculture has turned soil into dust

The 'Dust Bowl' era is when in US history storms were known as 'black blizzards', when soil turned into dust, and when more than 100,000,000 acres of land (400,000 km2) became a desert which sustained no life, neither of plants, nor of animals or humans.

For European readers, a million acres is the combined size of Germany, the Netherlands and twice the size of Luxembourg. The dust storms – or black blizzards – began in the 1930s, and lasted almost a decade. The first strong winds blowing dust across the High Plains sprang up in the Dakotas. By the end of the 30s, the Dust Bowl extended south through Colorado and into Texas, with the worst affected areas being western Kansas and the Oklahoma Panhandle. An estimated 2.5 million people were forced to pack up their belongings and leave their homes in search of work and a new place to settle. Most of the refugees chose to go west on Route 66, hoping to get jobs in one of California's many orange plantations and fruit orchards. But fruits are seasonal, and the harvest period is short. In 'Grapes of Wrath', John Steinbeck described the poverty, hunger and utter misery that awaited those who had fled to California. Until today, Tom Joad and his family, the fictional characters Steinbeck describes, stand for the fate of the 'Oakies', the Dust Bowl refugees from the Southern Plains with the Oklahoma Panhandle at the center. The author, radio journalist and historian Studs Terkel later traveled through much of the south and west of the United States to talk to eye witnesses and record their stories. The oral histories he collected tell not only the story of the dust storms in Oklahoma and Kansas, but also that of lives marred by unemployment, hunger, and poverty. It was the fate of so many people during the 1930s, the years of the Great Depression that followed the world economic crisis. October 24th,1929 is still known as 'Black Thursday', the day the New York Stock Exchange went into freefall, leaving many investors both broke and desperate. 'Black Thursday' made it into European history books, too, as it marked the beginning of the World Economic Crisis. In Germany it triggered not just a phase of hyperinflation and mass unemployment, the 1930s also saw the rise of the Nazis and, at the end of the decade, the beginning of World War II. Given the turmoil of those years, it is unsurprising that few people in Europe were interested in what was happening in a remote part of the Great Plains. And who was to know that what seemed like extreme weather events actually had more to do with international politics and world trade than the climate?

In the United States, too, the Wall Street Crash, high unemployment, the political developments in Europe, and the start of the war made for more frequent headlines than dust storms in rural Texas, Kansas and Oklahoma.

In the decades that followed, the civil rights movement, the Cold War, the Vietnam War, anti-war protests, and the social changes they ushered in left few people with a desire to look back at the Dust Bowl era. It was New York Times reporter Timothy Egan who, in 2002, started researching this particular phase of US history, just in time to interview men and women who had lived through the dust storms or had played a part in dealing with the aftermath of this decade-long disaster. In 2006, he published 'The Worst Hard Time. The Untold Story of Those Who Survived the Great American Dust Bowl'[1]. 'American meteorologists rated the Dust Bowl the number one weather event of the twentieth century. And as they go over the scars of the land, historians say it was the nation's worst prolonged environmental disaster', he writes in the introduction[2]. An environmental catastrophe that is about to repeat itself in the 21st century. Not only have dust storms been regularly occurring in Midwestern States, since 2011 they have been making headlines in Germany, too. On a Saturday afternoon, motorists driving on a highway in the

---

1 Timothy Egan: The Worst Hard Time. The untold story of those who survived the Great American Dust Bowl. Mariner Books 2006
2 Egan, page 10

east German state of Mecklenburg-West Pomerania suddenly and unexpectedly found themselves shrouded in almost complete darkness. The resulting mass pileup left several people dead and dozens injured. These days, traffic alerts warning of dust storms are no longer uncommon. The reason – then and now – comes from extreme drought conditions in rural areas with large, conventionally managed farming operations. In a news story published on April 9th, 2019, the German news agency dpa cites Mecklenburg-West Pomerania's agricultural secretary, Till Backhaus: 'At danger are all dried out, fallow fields as well as fields that have been freshly plowed with a dried-out surface.' Backhaus appealed to farmers: 'I urge everyone in the affected areas to adapt their practices to the present danger."[3]

The dust clouds blowing from fields in east Germany in no way compare with the intensity of the dust storms during the Dust Bowl era, but today, almost 100 years later, a look back at the 'dirty thirties' of the 20th century shows that the conditions that led to millions of tons of fertile soil being carried away by the wind are about to prevail once again.

The first white settlers arrived on the American east coast and stayed there. Explorers ventured deep into the North

[3] https://www.t-online.de/nachrichten/panorama/buntes-kurioses/id_85553036/mecklenburg-vorpommern-staubstuerme-drohen-die-regierung-bittet-bauern-auf-die-duerre-zu-reagieren.html

I'm about 5'7" in boots. Prairie grass roots would reach above my head.

American continent but settlers were slow to follow. That changed dramatically at the end of the 19th century when more and more railroad lines were built, making it easier to move out west. But the High Plains, the vast expanse of prairie land east of the Rocky Mountains, remained mostly untouched. Agricultural use started in the southwest. Ranchers kept cattle which cowboys would deliver to the nearest rail head during the annual cattle drive. From loading stations like Dodge City in Kansas, the animals were transported by rail to the slaughterhouses in Chicago. Life on the High Plains was difficult, for the ranchers as much as for the cattle. Drought, blizzards and snow storms, grass fires, hail, torrential rain, and tornadoes – life on the prairie meant dealing with it all. 'The weather might display seven different moods in a year, and six of them were life-threatening'[4], writes Egan. Initially, no one thought about growing crops on the High Plains. But many of the newly arriving settlers were Russian-Germans who had grown wheat around Odessa and in the coastal region of the Black Sea[5]. Initially, most just grew some wheat for their own needs, but then world politics and the intervention of the US government overnight started a wheat boom that can only be compared to the California Gold Rush. In 1910, a bushel of wheat still sold at eight US cents. 'Five years later, with world grain supplies pinched by the Great War, the price had more than doubled. Farmers increased production by 50 percent. When the Turkish navy blocked the Dardanelles, they unexpectedly did the dryland wheat farmers a favor that no one could have imagined. Europe had relied on Russia for export grain. With Russian shipments blocked, the United States stepped in and issued a proclamation to the plains: plant more wheat to win the war. And for the first time, the government guaranteed the price, at two dollars a bushel, through the war (…). Wheat was no longer a staple of a small family farmer but a commodity with a price guarantee and a global market'[6]. Farmers heard the message and started planting wheat: about 45 million acres (18 million hectares) in 1917 and more than 75 million acres (30 million hectares) two years later. To ramp up wheat production to such a scale, additional cropland was needed. It wasn't only the immigrants who had settled in the southern part of the High Plains that bought or rented land and invested in plows and harvesters. There were also the 'suitcase farmers' who had no intention of ever settling there. They wanted only to rent out a tractor and a piece of ground for a few days, drop some winter wheat into the fresh-turned fold, and come back next summer for the payoff'[7], writes Egan. He tells the story of one such suitcase farmer who came to southeast Kansas

---

4 Egan, page 22
5 More on that in chapter 7
6 Egan, page 42/43
7 Egan, page 50

in 1921 to plow 32,000 acres (13,000 hectares), only to cover twice as much four years later. Banks set a vicious circle in motion by advertising cheap loans and repayment terms for the acquisition of machines: to service their debts, farmers needed to grow more. They bought or rented additional land, most of it still covered by prairie grasses. In order to work it, they needed more and bigger machines. Egan reckons that between 1925 and 1930, around 5.2 million acres (two million hectares) of prairie were plowed up; that's in addition to the nearly 20 million acres (8 million hectares) of grassland farmers had converted into cropland earlier[8].

In 1929, Russia resumed exporting wheat again, meaning there was suddenly a surplus in Europe, while in the US unsellable wheat started to fill up grain bins until there was no storage space left. There was no option but to pile the wheat out in the open. The collapse of the grain market coincided with the 1929 Wall Street stock market crash. With the beginning of the wheat boom, many farmers had become investors, not in stocks traded in Wall Street, but in commodities. With the prospect of record yields that might be sold at record prices, they had become used to taking out huge loans – what better way to make up for this year's losses than expand next year? But in 1930, the world economic crisis led to mass unemployment, not just in Europe, but in America, too. Many people were so poor that they couldn't afford food, let alone buy any kind of consumer goods. 1930 was also the year in which farmers harvested a record amount of wheat, which led to a huge surplus, and despite prices tumbling, much of the wheat could not be sold at all.

By the fall of 1930, the 'suitcase farmers' had disappeared, says Egan. 'They had no sooner plowed up several million acres than they walked away, leaving the land stripped, not even planted in wheat. Just naked, exposed to the wind'[9]. In 1932, almost a third of all farms on the High Plains were so far in arrears with bank payments and taxes that foreclosure seemed imminent. More and more families just packed up and left in search of work. What remained were empty farmsteads and bare fields – in a region where there's almost always a strong wind blowing. At a windspeed of 30mph (50kmh), bare soil starts to move, and once the wind reaches a strength of 35 to 50mph (60 to 80kmh), dust storms begin to develop. With the High Plains counting as one of the cold, semi-arid regions of the world, water is scarce. The 1930s turned out to be a decade with below-average rainfall and above-average temperatures, two factors that boosted soil erosion. During the winter of 1932, dust storms started to occur regularly. What looked like the

---

8   Egan, page 58
9   Egan, page 80

dark clouds of a thunderstorm on the horizon were dust clouds which blew across the land with the force of a blizzard. Old black and white photos taken during the 1930s show houses half buried under dunes of sand, fences where only the top of the fence posts are still visible, cars and machines stuck in dust like in a snow drift. But it's the eyewitness accounts that really convey what human beings and animals went through while the dust storms were raging. Those experiences, in particular what agricultural communities had to endure, still have an impact on farmers today.

Abandoned farmstead in Kansas

One of these farmers is Michael Thompson – of whom you will hear a lot more in Chapter 9. At the end of our visit, he gave me a book. 'Farming the Dust Bowl' by Lawrence Svobida, a farmer from southwest Kansas, who originally came from Nebraska. In 1929, at the age of just 21, he moved to Meade County in southwestern Kansas and began farming. His first harvest was destroyed by hail. In the nine years that followed he battled dust storms, heat and drought until he gave up the farm in 1939. Exhausted, ill and out of money he moved in with friends. Eventually, he began writing down his recollection of those traumatic years in a spiral bound notebook[10]. 'Farming the Dust Bowl' was published in

---

10 I could not find much about Svobida's life. The information I do have comes from Hoehnle, Peter: "Beyond the Dust Bowl: Lawrence Svobida, 1908-1984." Agricultural History vol. 75, no. 3 (2001): 271-78

1940 by a small publishing house in Idaho. It was Svobida's second attempt to get his story into print. The first hand-written manuscript was returned by the editor with a note saying it was 'illegible'. Svobida died in 1984 in Fort Smith, Arkansas, a small town on the banks of the Arkansas River. Shortly before his death he wrote: 'I would not trade this area which is all timber, grass and cattle and mild winters for all of the high plains'[11]. Svobida's book was out of print for a long time. In 1986, it was republished by University Press of Kansas where it is still available today.

Usually, the storms on the High Plains started in February. Svobida writes about 1933: 'With the gales came the dust. Sometimes it was so thick that it completely hid the sun. Visibility ranged from nothing to fifty feet, the former when the eyes were filled with dirt which could not be avoided, even with goggles. When the dust is so fine that it will even penetrate the works of fine watches and stop them, there is no way of controlling it. During a gale the dust would sift into the houses through crevices around the doors and windows, eventually to lie an inch or more deep all over the floors, and on tables, chairs, and beds. Cleaning up a house after a dust storm is no picnic. Disturb the dust and it flies up. Choked and gagging, the cleaner has to run outside to breathe. I speak from personal experience. Often I have used a scoop shovel to remove the great quantities from my house'[12].

Sealed cans are the only containers dust cannot penetrate, writes Svobida. He found it impossible to eat anything during a storm without swallowing dust, while water in a bucket turned to thin mud. After each dust storm, farmers found dead animals. Even where cattle still had some grasses to graze, they often starved. Timothy Egan reports that a veterinarian who examined a dead cow found that her stomach was totally filled with compacted dust that had formed a huge, hard lump, making feed intake and digestion utterly impossible. Svobida writes about cows with teeth 'sanded down' to the palate plate because their feed contained so much dust.

In humans, the dust led to respiratory problems and diseases, and physicians saw cases of silicosis as severe as those found in miners who had spent all their working life underground.

---

11 Hoehnle page 276
12 Laurence Svobida; Farming the Dust Bowl, University Press of Kansas, 1986, page 62/63

One of the worst dust storms of the decade began on April 14th, 1935, a day that became known as 'Black Sunday'. The weather front was some 200 miles (over 300km) wide and moved with a speed of up to a 100mph (160kmh). 'Only those who have been caught out in a "black blizzard" can have more than a faint conception of its terrors', writes Lawrence Svobida. 'When the soil has become finely pulverized by too much working over, by the action of water followed by wind, or, particularly, when the surface is blow dirt from a previous storm, the dust begins to blow with only a slight breeze. As it continues to rise into the air it becomes thicker and thicker, obscuring the landscape and continuing to grow in density until vision is reduced to a thousand yards, or less. If this is to be a real dust storm, a typical black blizzard of the Dust Bowl, the wind increases its velocity until it is blowing at forty to fifty miles an hour. Soon everything is moving – the land is blowing, both farm land and pasture alike. The fine dirt is sweeping along at express-train speed, and when the very sun is blotted out, visibility is reduced to some fifty feet; or perhaps you cannot see at all, because the dust has blinded you (…) Birds fly in terror before the storm, and only those that are strong of wing may escape. The smaller birds fly until they are exhausted, then fall to ground, to share the fate of thousands of jack rabbits which perish from suffocation. Human beings run for their lives, if there is any safe place within reach. Some run anyway, well knowing that unless shelter is reached, they may be victims of the same fate that overtakes the birds and the jack rabbits'[13].

In the early 1930s, when the first dust storms swept across Kansas, Texas and Oklahoma, farmers like Lawrence Svobida tried to protect their fields by deep plowing across the direction of the wind. The deep soil layers which now were atop still held a little bit of moisture and usually had some tilth – such soil wasn't as easily picked up by the wind and if it did blow, it mostly ended up in one of the field's deep furrows. The problem was that farmers not only plowed deeper and deeper but they also did so after every storm. Because the fields mostly lay fallow, the soil contained little soil organic matter and practically no soil life – living soil had turned into dirt and dust. Even the 'lucky' farmers who hadn't lost all topsoil from their fields and were able to sow some winter wheat rarely had a harvest: in spring, the storms usually blew across enough dust from neighboring fields to suffocate the tender, young wheat plants that had made it thus far. In southwestern Kansas, the layer of fertile topsoil is not thicker than 4 inches (10cm), says Lawrence Svobida, beneath lies mostly inorganic subsoil. 'There is a rock in a neighbor's field, which used occasionally to be struck by the plow,

---

13 Svobida page 123f

because it lay beneath the surface. That rock now stands four feet above the surface, and the land is still blowing with every wind'.[14]

At the end of his book, Svobida concludes that large swathes of the Great Plains will never recover from the 'Dust Bowl' and remain an infertile desert. That this bleak forecast didn't become reality is mostly down to one man, yet to this day his name is not well-known, even in the US. Hugh Bennett grew up on a farm in North Carolina. He attended university and focused on different methods of soil cultivation and tillage. Immediately after graduation, he got a job as a surveyor and was part of a government team of scientists to conduct the first ever nation-wide soil survey. Once the work on the soil register was finished he returned home to take care of the farm. Soil remained his passion. Bennett continued to work as a consultant, and from the late 1920s onwards he began warning publicly about soil degradation and erosion: on the Great Plains the government was encouraging farmers to work against nature, and that would have dire consequences.

Franklin D. Roosevelt was elected president in the crisis year of 1933. As a measure to alleviate poverty and hunger in rural communities, his government bought up surplus grain and distributed it amongst families in need. But Roosevelt also looked for ways to change the conditions that had led to the crisis – and that's how Hugh Bennett got an invitation to the White House. He explained to Roosevelt what damages wind can cause in an arid, flat expanse like the High Plains when the soil isn't protected by prairie grasses but has been plowed up and left fallow. Bennett predicted that the frequency and intensity of the dust storms would only increase. Roosevelt decided to create a new agency within the Department of the Interior that was to look into how soil could be stabilized on a grand scale. He offered Bennett the post of director. The plan Bennett came up with was nothing short of an attempt to rewrite the history of agriculture, says Timothy Egan in The Worst Hard Time. 'One idea was to put new growth on the bald grasslands, a restoration project that had never been tried before on such a scale. His other idea was to get individual farmers to break down their barriers of property and think beyond their fence lines. It wasn't enough for one farmer to practice soil conservation if his neighbor's land was blowing'[15].

The agency tasked with regenerating soils was the newly created Civilian Conservation Corps. Field staff set up demonstration plots to show how vegetation could be brought back. To break the power of the wind millions of trees were to be planted – an idea Roosevelt particularly liked and supported. But in 1933 Bennett's annual budget amounted to just five million dollars – nowhere near

---
14 Svobida page 255
15 Egan page 159

enough to put plans into practice that aimed to stop erosion and permanently stabilize soils. 'One man cannot stop the soil from blowing but one man can start it'[16], was Bennett's mantra. He needed money, and to get it he needed to convince Congress to grant it. In April of 1935, Bennett was scheduled to appear before a Senate committee. From weather reports, he knew about the 'black blizzard' that was developing over the Great Plains; the forecasters predicted the storm would be so huge as to be felt on the East Coast, too. A year earlier, a storm of similar magnitude had darkened the skies over New York City and Washington D.C., and dust from the High Plains had been blown onto the decks of ships in the Atlantic, some hundred miles off the US coast. Bennett asked the Senate to postpone the hearing for a few days, and it was rescheduled for April 19th. On Friday afternoon, five days after 'Black Sunday' on the High Plains, Bennett explained his plans for soil protection and the prevention of soil erosion to a group of bored senators. Timothy Egan gives a vivid description of what happened next: 'A senator who had been gazing out the window interrupted Bennett. "It's getting dark outside." The senators went to the window. Early afternoon in mid-April, and it was getting dark. The sun over the Senate Office Building vanished. The air took on a copper hue as light filtered through the flurry of dust. For the second time in two years, soil from the southern plains fell on the capital. (…) "This, gentlemen, is what I'm talking about," said Bennett. "There goes Oklahoma."'[17]

Within days, Bennett got the money to finance his project, and the agency for soil conservation was made permanent. Congress passed the 'Soil Conservation Act', which Roosevelt signed into law at the end of April 1935.

Bennett and his staff began identifying grass varieties that would quickly take root in dusty, degraded soil with a tendency to move and shift like sand dunes. They decided which areas should be given priority, and the US government bought up more than 10 million acres of farmland destroyed by wind erosion with the aim to convert it back into permanent grassland.

In Kansas and other regions, crop cultivation would continue. In those areas, millions of trees were to be planted as 'shelter belts'. They would not be strong enough to stop a dust storm, but might break the force of the wind somewhat and give some protection. A team of 11 men could plant 6,000 trees in a day, writes Egan, and until 1938 almost 40 million saplings were planted. Most were spindly 'sticks in the ground', but the hope was that over the years, they would grow into thick hedges – forming 3,600 miles of shelter belts.

---

16 Egan page 226
17 Egan page 228

New settlers in a dugout – exhibit at the Homestead National Monument, NE

At the beginning of the new decade, rainfall patterns seemed to normalize; the extreme drought of the 1930s had ended. After the end of World War II and with the invention of center pivot irrigation, more and more land in the west could again be used to grow wheat, corn and sugar beets. The pumps sucking up water from the Ogallala Aquifer run day and night, and throughout the growing season. Today, a third of all irrigated cropland in the US is on the High Plains, irrigated with water from the Ogallala Aquifer. Aquifers are porous rock formations that store water for millennia. The Ogallala Aquifer stretches from South Dakota to northern Texas, but even such a huge water storage system can become exhausted – for decades more water has been pumped out than could be replenished through snow-melt and rain. It is not a question if, but when farmers will no longer have water to irrigate their crops. Then, much of the land on the High Plains will once again either be covered by prairie grasses or farmers will decide to continue plowing, creating the conditions for a new Dust Bowl age.

Roosevelt's shelter belts have all but vanished. Most of the hedges were ripped out to make space for a few additional rows of corn or soy, and operating large drills and combines, too, is easier if one does not have to watch out for trees and shrubs.

Without human interference, the dust storms of the 1930s would not have happened. Converting millions of acres of prairie into cropland was the trigger, but two other factors made a bad situation worse: heat and drought led to decreasing yields and increased the power and damage the 'black blizzards' could unleash. In 2016, scientists at the University of Chicago created a simulation: what would yields be like today if conditions such as temperature and rainfall were equivalent to those in the Dust Bowl era?[18] Michael Glotter and Joshua Elliott first looked at the impact of rising temperatures and concluded that yield losses until the middle of the 21st century would be similar to those during the 1930s, even if there was enough rain: yields for corn and soy would drop by 40 percent, and those for wheat by 30 percent. 'The harm would be 50 percent worse than the 2012 drought, which caused nearly $100 billion of damage to the U.S. economy'[19], according to the findings of Glotter and Elliott. If rising temperatures were to coincide with drought, yields in the Midwest could be down by 80 percent by 2050. 'We expected to find the system much more resilient because 30 percent of production is now irrigated in the United States, and we've abandoned corn production in more severely drought-stricken places such as Oklahoma and west Texas. But we found the opposite: the system was just as sensitive to drought and heat as it was in the 1930s'[20], says Joshua Elliot. Agricultural practices have changed tremendously in the 90 years since the Dust Bowl, says Michael Glotter: 'But many technological and geographical shifts were intended to optimize average yield instead of resilience to severe weather, leaving many staple crops vulnerable to seasons of unusually low precipitation and/or high temperatures'[21].

Four agrochemical companies, Bayer-Monsanto, Dow-DuPont (now Corteva), Syngenta (bought by ChemChina) and BASF sell 60 percent of the world's seeds. In 2008, Bayer-Monsanto supplied 80 percent of genetically engineered (GE) corn grown in the US, 86 percent of GE soy, and 92 percent of GE cotton. The percentage keeps increasing. The 'Big Four' have one thing in common: test fields on Hawai'i.

---

18 https://www.nature.com/articles/nplants2016193
19 https://news.uchicago.edu/story/dust-bowl-would-devastate-todays-crops-study-finds
20 Loc. cit. https://news.uchicago.edu/story/dust-bowl-would-devastate-todays-crops-study-finds
21 Loc. cit. https://news.uchicago.edu/story/dust-bowl-would-devastate-todays-crops-study-finds

Hawai'i – Industrial agriculture's ground zero

Of course US citizens do know where Hawai'i is situated as it is one of the nation's 50 states. For Europe though, Hawai'i is on the other side of the globe, a chain of tiny dots in a sea of blue. Hawai'i consists of a group of islands some 2,500 miles (almost 4,000km) west of the US mainland. The most southern island is Hawai'i Island. It is the largest one, and is usually just referred to as 'Big Island'. At the other end of the island chain, 300 miles to the northwest, lies Kauai. Oahu with the capital Honolulu, Maui, Molokai, and some tiny, uninhabited islands sit in between. Steep volcanic slopes characterize the islands; some have been dormant for centuries while others are still active. In April 2018, an earthquake shook the Big Island; cracks and fissures appeared on and around Kilauea, one of its volcanoes, oozing and spewing molten lava which devastated several neighborhoods. Downwind towns and villages were hit by smog and at times poisonous gases. The eruption of Kilauea was so powerful that it could be clearly seen from space.

The mountain ranges on the islands form a watershed, making for a dry westside and an eastern side with plenty of rain. Ferns grow on the steep slopes of the volcanoes, palm trees fringe the sandy beaches, and coral reefs break the force of the waves, endlessly rolling in from the Pacific. About 1.4 million people live on Hawai'i but add to that ten million tourists per year who come to enjoy a few days in a tropical island paradise.

What they don't see is the literal flip side of Hawai'i, the part of the island where agrochemical companies have their office buildings, laboratories, and test fields. Bayer[1]Monsanto, DuPont/Pioneer, Syngenta, BASF, Dow AgroScience – all of them have either a direct presence on Hawai'i or work with a partner organization. The seeds that are developed and tested on Hawai'i are what farmers will plant on millions of hectares of land at the start of the next growing season – in the Midwest of the US, in South America, in Asia, and in Europe. For the seed production on Hawai'i, all means of industrial agriculture are being employed – from genetic modification to the use of chemical fertilizers and every herbicide, pesticide and fungicide known to man. What happens on the test plots on Hawai'i is industrial agriculture on steroids.

The forms of extreme industrial agriculture practiced on Hawai'i have consequences

---

[1] Bayer acquired Monsanto in June of 2018, after the events described in this chapter. Therefore I will refer to Monsanto, rather than to Bayer-Monsanto.

– for people's health, the environment, biodiversity and soil quality. Having a close look at what's happening on Hawai'i is like a trip into the near future: this could become the 'new normal' wherever industrial agricultural practices are not impeded by regulations and laws and agrochemical companies are not held accountable for the harm caused by the products they sell.

Marghee Maupin, family nurse practitioner on Kauai

From the veranda of her house, Marghee Maupin enjoys a wonderful view over a lush, forested valley. It's the end of October, and it is hot, the tropical rain showers bring little relief, and if it weren't for the constant slight breeze coming in from the sea, Kauai would feel like a giant hothouse with the roof removed. It's because of this climate that Kauai, the fourth largest and oldest of the Hawai'ian islands, is also called the Garden Island.

Maupin is a community nurse. Since 2005, she has run a small mobile health clinic together with her husband, offering general healthcare and providing care for patients with chronic conditions like diabetes, high blood pressure, kidney problems and respiratory diseases. A lot of patients show up with acute health issues – shortness of breath, difficulty concentrating, headaches, rashes, and severe nose bleeds. In particular, children sometimes wake up with a pillow soaked in blood. And Maupin cares for cancer patients. Many indigenous Hawai'ians[2] still live in large, joint families. 'I know one family here where 35 family members have been diagnosed with cancer. The youngest is a 14-year old boy with testicular cancer,' Maupin says.

The beautiful side of Hawai'i tourists come to see

Marghee Maupin is in no doubt that many of the symptoms she sees in her patients are caused by the intensive and continuous use of pesticides on the island. Until 2015 Maupin worked at a clinic in Waimea, a town with about 9,000 inhabitants in the southwest of Kauai. Very few of the 1.2 million tourists who visit Kauai every year ever set foot in Waimea – there is no need, for the road leading to the 'Waimea Canyon State Park' veers off to the north just before one gets into town. Many mountain peaks on Kauai rise well over 3,000 ft (1,000m) above sea level. The panorama from the trails and the viewpoints in the park are stunning; the slopes consist of hundreds of ridges, each dropping off almost vertically on either side, making the mountain sides appear covered by an elaborately folded emerald green robe, fringed by the white crests of the Pacific waves rolling in. Low-

---

2 Polynesian sea farers settled on Hawai'i around 400 AD they reached the islands from what are now the French Polynesian Marquasas islands, more than 2,000 miles (3,500 km) away. Over the centuries, many ethnicities settled on Hawai'i, but when mention is made of indigenous Hawai'ians, this refers to those of Polynesian descent.

hanging, wet clouds suddenly sweep across the viewpoint and, within seconds, you are shrouded in fog and then drenched by a cold rain shower. A few minutes later the sun is out again and a double rainbow spans across the valley. Most tourists take time for yet another selfie, and then it's time to move on to the next beauty spot.

Test fields on Kauai – visitors not welcome

The next day we pass the junction that leads to the State Park and continue on the coastal road into Waimea. A large billboard outside the school advertises the play to be performed on Halloween, 'Nightmare in Waimea'. Fields border the school ground, and leaving the town, there are fields on both sides of the road; the crops have been harvested, and some fields have been freshly plowed. Then a large complex of one-story buildings comes into view – that's the headquarters of Hartung Brothers, Inc., Syngenta's partner organization on Hawai'i. In the summer of 2017 ChemChina took over Syngenta, and in the course of the merger Syngenta sold all its assets on Hawai'i to Hartung Brothers, Inc., a US agribusiness operating out of Wisconsin which has been working with Syngenta for a long time. According to media reports[3], Hartung Brothers, Inc. continues Syngenta's operation on Hawai'i in its entirety. On Kauai, the reason for the sale seems obvious to everyone we talk to: the road continues only for a few more

---

3   http://www.thegardenisland.com/2017/05/12/business/syngenta-sells-hawaii-sites/

miles and abruptly ends at the locked and secured gates of a military base – one of 142 such compounds and installations on Hawai'i. After the Syngenta-ChemChina merger, Chinese staff would have had unfettered access to the test fields situated next to the military base. With Syngenta selling off its assets to Hartung Brothers, Inc. the work can continue as before, but Chinese visitors have no access.

The black top road ends at the gates of the military base, but a dirt road continues for another mile along the coast until it becomes a barely discernible footpath. There is nothing here but wasteland. The only structure is a deserted shelter for farm workers with a few broken metal seats. Because of the recent rain, the red dust that blows from the fields has formed a thin layer of mud. The only other signs of human presence are the markers, plaques and signposts: No access! Trespassers will be prosecuted! Strictly no entry! The signs do not mark the borders of ordinary fields. We are standing next to some of the test fields that agrochemical companies maintain not only on Kauai, but on all Hawai'ian islands. It takes a fair amount of research to find out where exactly some of these test sites are located. They can be in a barely accessible area beyond a military base, secured by warning signs and legal notices, others are surrounded by high fences, with yet more only visible from the air: corn growing on tiny plots in the middle of what otherwise is a sugar cane plantation. Local residents get a different type of proof when test fields are nearby: when pesticides are being used you can smell them, says Marghee Maupin, the community nurse. 'Suddenly the air smells of bubble gum or there is a weird artificial flower scent,'. Crop-spraying plans are the industries' well-guarded secret. There is no warning for people who live or work downwind, for school kids who might be outside on a break. The companies give no information when fields will be sprayed, nor what chemicals will be used.

The Center for Food Safety, CFS, is a non-profit organization, advocating human health and environmental protection by curbing the use of agrochemicals and food production technologies like genetic engineering of seeds, while promoting organic and other forms of sustainable agriculture. In 2015, CFS published a comprehensive report titled 'Pesticides in Paradise. Hawai'i's health & environment at risk'[4].

The school in Waimea is situated directly next to several test fields. At least three times students became seriously ill after pesticides were used nearby. 'In a 2008 episode, 60 children and at least two teachers experienced headache, dizziness,

---
4 https://www.centerforfoodsafety.org/issues/3859/hawaii-center-for-food-safety/reports/3901/pesticides-in-paradise-hawaiis-health-and-environment-at-risk

nausea, and/or vomiting; ten or more children were treated at an emergency room; several were put on a nebulizer to relieve respiratory distress; and one was given an anti-vomiting medication intravenously. A teacher who was also affected firmly rejected the explanation given by Hawai'i officials and Syngenta that "stinkweed" was the culprit, saying that she was familiar with stinkweed's odor and that this was not the cause.'[5] Marghee Maupin was working in Waimea at the time and remembers the incident well. A residue analysis later confirmed that the symptoms were caused by chlorpyrifos, an insecticide that belongs to the organophosphates group, which affects the central nervous system. Germany banned the sale of pesticides containing chlorpyrifos in 2009. In the US, the Obama administration had prepared a chlorpyrifos ban which was to come into effect in early 2017, but Scott Pruitt, the new head of the EPA appointed by President Donald Trump, put it on hold. Today, several US states have passed legislation to phase out the use of chlorpyrifos, while environmental groups have taken the fight for a national ban and against the EPA to the courts where it is making slow progress.

No subtle warning

It is almost impossible to find out which agrochemicals are being used. Agrichemical companies are under no obligation to disclose what they spray – unless they

---

5 Lit.cit. page 38

are forced to do so through a court order. That happened in a court case on Kauai. 150 families living in close proximity to DuPont/Pioneer test fields sued for compensation, arguing that the pesticide-laden dust blowing across from the fields was harming their health and the environment. In court, several DuPont/Pioneer staff were questioned under oath, and it could be established that the company regularly used 80 different pesticides and adjuvants (chemical agents that make the active ingredient more effective)[6]. The court ordered DuPont/Pioneer to pay compensation for the damage pesticide dust was causing to the plaintiffs' houses, but in order to get any money, they had to sign a waiver: in regard to health issues they would at no point in the future make any claims against DuPont/Pioneer. Nevertheless, the court case was very important because it officially established the frequency with which agrichemical companies use pesticides, says Marghee Maupin: test fields are sprayed, on average, on 260 days in a year, and up to 16 times a day.

It is not easy to find work on Hawai'i. The big employers are the tourism industry, the military, and the agrichemical companies. On the west side of Kauai, DuPont/Pioneer and Syngenta (now Hartung Brothers, Inc.) offer work; almost everyone has a family member, a neighbor or friend who has a job with one of the two companies. Yet there is general awareness that the agrichemical companies apply large amounts of pesticides and spray on a daily basis. Of course, everyone has heard about the children at the Waimea school who suddenly fell violently ill. One knows that farm workers employed by Syngenta had to be rushed to A & E because they had accidentally walked into a field that had just been sprayed. Everyone knows the symptoms because most people have experienced them at some point. The dilemma is obvious: what is more important, one's health or having a job and money to live? To protest against the pesticide use means risking one's own job or that of a family member.

On a rainy morning in October, we meet Gary Hooser in Koloa, a small tourist town in the south of Kauai. You can rent surf boards, book diving excursions or buy a beach outfit. There are coffee shops, high-end restaurants and quiet side streets with nice houses surrounded by well-maintained gardens. With his thick, greying hair and distinctive profile, Hooser reminds me of the late Senator, Edward Kennedy, except that he isn't wearing a suit and tie but an immaculately ironed Hawaiian shirt which has become something of a trademark. For four years he was a member of Kauai's county council. From 2002 to 2010 he was

---

6   http://gmwatch.org/en/news/archive/2015-articles/16154-dupont-pioneer-found-guilty-of-dust-pollution-and-trespass-in-kauai

the majority leader of the Hawaiian State Senate. After an unsuccessful bid to become governor, Hooser returned to Kauai and once again ran for the county council. 'I don't seek out challenges but the challenges find me,' says Hooser. During the election campaign in particular, young voters kept coming up to him, wanting to discuss the agrichemical companies and the pesticide use on the island, remembers Hooser. 'There was anecdotal evidence of people getting sick. There were the dead sea urchins that were found near the west coast where the companies have fields close to the shore,' he recalls. 'People said they thought there were a lot more cancer cases on the west side. They asked whether their drinking water might be contaminated with pesticides. In particular young voters said I needed to do something.' Hooser was elected and decided on a pragmatic approach: 'I am just one person, we can't just get the companies off the island, there are legal issues, but I can look into it. I started to ask questions and found that the answers I got were less than truthful. So, I asked more questions. I met with representatives of the ag companies and said: people are concerned, please help me out here. We are not after trade secrets, we are not wanting to get you off the island, we would just like to know what you are spraying.'

The answer he finally got was a two-page letter in which the companies politely stated they were not going to disclose any information. They said their use of pesticides was in line with common practice in agriculture elsewhere, guidelines for dosage and application were being strictly adhered to. For some herbicides, that is totally impossible, says Hooser, Dicamba, for example. is volatile and prone to drift. Label instructions state that it should only be sprayed at a wind speed of 10mph or less, conditions that practically don't exist on Kauai because there is always some wind. So, the companies went to the Department of Agriculture in Honolulu and asked for an exemption. 'I was shocked to find that the department didn't even ask how much these companies used. And the companies clearly lied to me'. Only for one group of pesticides do farmers and agricultural companies have to disclose the amounts they use on a monthly basis. These are so-called Restricted Use Pesticides or RUPs, which are particularly toxic and can only be used by licensed applicators and under strict conditions. Atrazine, chlorpyrifos and paraquat are all RUPs, and they are banned from use in some or all EU countries. Gary Hooser and environmental organizations like CFS put a lot of effort into finding the government websites that list the special permissions given to agrichemical companies and other applicators. They were able to establish that in one year alone, 18 tons of 23 different RUPs were sprayed on Kauai. But for most pesticides, there are no restrictions. For example, nobody records how much glyphosate is being used. And the agrichemical companies criticized

Hooser's calculation. The active ingredient, for example atrazine, made up only a small part of any commercially available herbicide, they argued. Therefore, the use of pesticides on Kauai amounted to a few kilograms and definitely not tons. All he ever got from the industry were half-truths, lies and stone walling, says Gary Hooser. But he continued to avoid outright confrontation. Together with the other members of the county council he drafted what later became Bill 2491, stipulating that agrichemical companies in future should declare which pesticides they used. There would have to be a buffer zone between test fields and nearby schools and residential areas, and the bill proposed a moratorium: the companies should not be allowed to expand until an environmental impact study was done. 'The agrichemical companies reacted as if the sky had fallen in. The degree of emotion and intensity went through the roof from day one,' remembers Hooser. Suddenly nobody on Kauai talked about anything but the proposed bill. 2,500 people joined a protest march in Kauai's capital, Lihue. A public hearing on the effects of pesticide use on Kauai took until well into the night because there were so many people who wanted to testify. The agrichemical companies or the 'chemical cartel', as Gary Hooser now called them, did everything to stop Bill 2491 from being passed. At the time, MiKey Boudreaux was a volunteer for the non-governmental organization Hawai'i SEED, and in June of 2013, it became a full-time job. There were only 70 seats for visitors in the county council. The agrichemical companies paid homeless people to stand in line hours before visitors would be admitted to the chamber; once the debate was about to start company staff took their places and filled the visitor ranks[7]. 'On the day of the vote, we spent the whole night outside of the building,' says Boudreaux. 'It was an inspiration to see the campaigners. They slept on the sidewalk because they believed in what we were doing,' says Gary Hooser, for whom it was an incredibly emotional time, too. He was maligned by the industry and attacked by members of the community. 'The industry is very well connected in this very small community. People in the mayor's office or on the board of the hospital, they all have links to industry or spouses who work for the companies,' says Hooser, who was threatened and verbally attacked almost on a daily basis. On social media he and members of his family were threatened physically, too. In the county council the decision finally came down to one council member. 'It was 3am, everyone was so exhausted, I was in the council chamber, crying', says MiKey Boudreaux. 'outside, protestors were still shouting "pass the bill" so loud that the windows rattled. And then this one council woman voted "yes".' The joy of having won the vote and the bill being

---

7   A few days after my conversation with MiKey Boudreaux, Gary Hooser confirms that agrichemical companies paid homeless people to stand in line. He said company staff, too, were ordered to queue early. They were served breakfast by one of the company's food trucks.

passed lasted only for a few hours. The ag companies filed a law suit the very next day. 'They sued us for the right to poison the land around our schools and hospitals,' says Hooser, who wasn't at all surprised that the 'chemical cartel' should start legal proceedings. For him, the lightbulb moment came during one of the county council meetings: 'It was crazy. I was chairing the meeting, I was the right person do so, I had been in the senate, I wouldn't be intimidated. And then came this one telling moment when I asked the representative from BASF "Why are you fighting us so hard? We are not after any trade secrets, we just want to know what you are spraying so that people can be kept safe", and he replied "We are concerned that other communities might do the same thing".'

The BASF rep wasn't wrong in his assessment. 'The agrichemical companies were about to face the perfect storm,' says Ashley Lukens, director of the Center for Food Safety on Hawai'i. The problem for companies like Monsanto started with the GMO labelling campaign in California. In the US, almost all processed food contains genetically modified ingredients, in particular the ubiquitous high fructose corn syrup (HFCS) derived from genetically engineered corn. As in Europe, many customers in the US are very concerned about possible health risks of GMOs in food, but while in Europe genetically modified ingredients are not allowed[8], US consumers have no way of knowing what goes into the products they buy. This led to a nation-wide debate and in California to a state-wide campaign to make labelling a legal requirement. Food and agrichemical companies spent millions of dollars to undermine and stop this initiative.

At the same time, the agrichemical companies on Hawai'i struck lucky: sugar cane plantations had been unprofitable for a while. Now the last ones were about to close, which meant much needed agricultural land would become available.

Ashley Lukens mentions a third, seemingly unrelated but very important factor: the budding 'Hawai'ian Renaissance'. Many indigenous Hawai'ians began to rediscover and engage with cultural traditions: the Hawai'ian language, the spiritual meaning of Hawai'ian dance and music, and the traditional agricultural system which was adapted perfectly to the conditions on the islands and provided the islanders not just with food, but building material, fibers, dyes and medicinal herbs, too. (More of that in Chapter 5). One of the leading proponents of the Hawai'ian Renaissance is Walter Ritte. Born in 1945 on Maui, he has been living on the small, neighboring island of Molokai for many years. Monsanto has a large

---

8   Feed is the big loophole: GE corn and soy are permissible as feed. But consumer organizations have started various campaigns for GE-free dairy products, and meat and supermarkets have reacted. Most now have 'GMO-free' product lines, guaranteeing that the farmers producing labelled dairy, meat, chicken and eggs used only non-GE feed.

number of agricultural test sites on both islands. Pesticide use on Maui and Molokai is as intensive as on Kauai. Residents on Maui and Molokai therefore wholeheartedly supported the efforts on Kauai for more transparency as to what chemicals the 'chemical cartel' was using and implementing buffer zones for better protection of schools and residential areas. And they started wondering what form a similar protest could take on Maui. Walter Ritte and the well-known Hawai'ian guitarist Makana organized protest marches for which thousands turned up – and for the first time, indigenous Hawai'ians were amongst the demonstrators.

To understand why the participation of native Hawai'ians was a game changer one needs to take a look at Hawai'i's history. For centuries, the Polynesian seafarers who came to Hawai'i from other remote Pacific islands were the only settlers. The first European sailors did not arrive before the 18th century, followed by missionaries and American traders and entrepreneurs. From the middle of the 19th century, US companies began to establish large sugar cane plantations, followed a little later by pineapple plantations. Laborers were in short supply, so the fruit companies and sugar processors started to recruit workers from China, Japan, the Philippines, and Korea. They came in their thousands, bringing their families, ready to settle on Hawai'i. US economic interests had political consequences: the last Hawai'ian sovereign, Queen Liliuokalani, was disempowered and Hawai'i annexed by the United States. In 1959, Hawai'i became the 50th US state.

Flip-flops, shorts, and t-shirts are the right attire for Hawai'i's tropical climate. The beaches are palm fringed, the surf is second to none, it makes sense to have a nap during the hottest part of the day and make the most of balmy nights: Hawai'i had a lot to offer pensioners, surfers, hippies, and anyone else who thought life on the mainland was just too stuffy, regimented, and hectic. Many came and never left. Today, 28 percent of Hawai'i's residents are white, 37 percent are Asian and only 10 percent are indigenous Hawai'ians of Polynesian descent. But Hawai'i is anything but a melting pot in which ethnicity and social status cease to matter. In the fight against the agrichemical companies, the social fault lines turned out to be very important. Hawai'i's colonial history was shaped by a few big companies and their plantations. As a result, Hawai'ians of Polynesian descent remain a marginalized group and economically worse off. They are poorer than other Hawai'ians, they are less educated, and their rate of unemployment is higher than average. Statistics also show that indigenous Hawai'ians suffer more health problems, their living conditions are often poor, and they are more likely to live in close proximity to the test sites of the agrichemical companies.

MiKey Bourdreaux moved to Kauai from the US mainland. She and other activists worked hard to raise awareness. Their hope was that once Hawai'ians understood the dangers they faced through pesticide spraying they would join the protest movement. 'It was really difficult,' she says. 'Native Hawai'ians saw us hand out fliers, protest and queue to attend council meeting, and they saw it as something only white folks could do because they simply didn't have the time. I was often told: if you were working three jobs to feed your family you wouldn't have time to demonstrate either.'

'Haoles' is the term indigenous Hawai'ians use for the white Americans who have moved to the islands from the US mainland. It isn't a derogatory term but it expresses how little the migrants from the mainland know about Hawai'ian traditions. To indigenous Hawai'ians, calling someone 'Haoles' implies the arrogance and the casual way in which white people seem to assume they know what is best for Hawai'i and Hawai'ians.

The organizers of the anti-pesticide campaigns on Hawai'i are well aware of the difficult and often fraught relationship with indigenous Hawai'ians but there is little they can do about it. 'The only thing we Haoles can do is to always remember that we are Haoles' has become the motto many activists have adopted.

In this situation, Walter Ritte and other indigenous Hawai'ians supporting the anti-pesticide protests on Maui from the start was a huge morale booster. In addition, an experienced organizer and veteran environmental campaigner joined the movement. Autumn Ness grew up in California, but before she moved to Hawai'I, she had lived in Japan.

The 11th March 2011 was the last day of Ness's holiday. A passionate surfer, she was visiting friends on Hawai'i and making the most of the waves on Maui's north shore. She was on the beach when suddenly the sirens on the islands went off. Tsunami alerts are fairly common on Hawai'i. In every building situated within a few hundred yards of the beach there will be a map on display with escape routes marked in red and the estimated time needed to reach higher, safer ground. Back at her friends' house, Ness learnt about the tsunami that had hit Japan. It took her more than two hours until she was finally able to reach her husband by phone – he had been travelling and was uninjured, but his home town, where he and Autumn lived, had been completely destroyed by the tsunami. In the following days, the full extent of the devastation in Japan became apparent, not only coastal towns and villages had been destroyed, but there was also the danger of a complete meltdown at the Fukushima nuclear power station. When Autumn Ness returned to Japan, she began working as a translator for different international aid organizations, and she regularly published the data on radiation gathered by scientists in independent laboratories. The Japanese government insisted that life outside the exclusion zone was perfectly safe, but there were good reasons not to trust these assurances and few people did. Six months later, Ness found out she was pregnant and decided that it wasn't safe for her to continue working in Japan. She returned to the US, and in February of 2012, almost a year after the tsunami, she moved into a small house on the outskirts of the town of Kihei on Maui. Just a few blocks away, and separated only by a simple fence, was agricultural land. At the time, Ness didn't know that these weren't ordinary fields but a Monsanto test site. Only the red dust that kept blowing from the fields towards the residential area was irritating – there are strong trade winds on Maui that come from the mountains and blow towards the sea. During a chance conversation on the beach, she learnt that the fields behind her house belonged to Monsanto and that the red dust was laden with pesticides. She decided to investigate the matter. 'I have learnt my lesson in Japan, the government is not going to protect you, you need to fight this,' says Ness.

On a particularly hot day in 2012, she got up early to open the windows. In Maui's tropical climate, air-conditioning is not so much a luxury as a necessity, but running it throughout the hot season is very expensive, and many people just can't afford it. Opening the windows early in the day is their only chance to get some cool air into their homes. That day, Autumn Ness knew immediately that something was wrong. 'I had a metallic taste in my mouth and within a few hours I had an excruciating headache,' she remembers. On the same day she went to see her doctor and asked to be tested for pesticide exposure. She was told that a lot of people come in with similar complaints but cannot be tested because they don't know what they might have been exposed to. Ness was advised to contact the Monsanto helpline to find out what pesticides had been sprayed that day. Ness never managed to speak to anyone from Monsanto on the phone – all she ever got was automated recordings. So, she drove up to the gate closest to the Monsanto office building and demanded to speak to a member of staff, a futile attempt, no one would be available, not now and not later, she was told. Autumn Ness decided to start a public campaign for more information and a buffer zone that would give the residents at least some protection.

No one on Maui chooses to live right next to a Monsanto test site, but it is nearly impossible to find low rent accommodation on the island. Home owners prefer to rent out houses or flats on a weekly basis to well-heeled tourists. On average, Hawai'ians have to spend more than half of their monthly income on rent – for Germany the rule of thumb is that housing expenses should not rise above 30 percent of monthly income.

Ness started talking to her neighbors and found that many were suffering from symptoms like headaches and nose bleeds, and everybody was annoyed at the red dust.

Together with Ness we drive through the residential area in Kihei. The houses could be the backdrop of an advertisement campaign for Barbie accessories: over the years the red dust has dyed every white surface pink.

Not only residents living near the test sites physically experienced Monsanto's presence. And that's why we wanted to meet Alika Atay. He is a Hawai'ian of Polynesian descent, a farmer and since 2016 a member of the county council.

Farmer and Maui county council member Alika Atay

For us, the day had begun at 4 am, with a flight from Kauai to Maui, with a two-hour stop-over in Honolulu. Since the night before, a storm had been raging over Maui, its intensity taking even meteorologists by surprise. The whole island was without electricity, and until the lights came on again at the airport in Kahului, our plane remained on the ground in Honolulu. With a delay of an hour we did finally get to Maui. There were still sudden gusts of high wind, strong enough to make our small rental car rattle and shake. Occasionally, it rained so heavily that sitting in the car felt more like being in a gigantic carwash than driving on a highway. It didn't help that the traffic lights still weren't working. Nevertheless, we did manage to arrive at the council building on time – to find it locked and Alika Atay nowhere in sight. When I finally reached him on his mobile, he confirmed that Maui was basically still closed down. No one was working, schools remained shut and as there was still no electricity at his house he and his family were hanging out at a sports bar watching THE game. Even for a European journalist, there seemed to be no excuse for being unaware of the fact that right now the decisive game in the Baseball World Series was taking place. We were welcome to join him and talk at the bar. (For fellow Europeans unfamiliar with baseball –as in cricket, there are longish periods in the game where nothing happens, followed by a sudden fast and furious bout of action – not ideal conditions to conduct an interview, but not impossible, either.)

Ten minutes later, we found the sports bar and located the Atay family. The Dodgers were in the lead, and Alika Atay has time to explain the concept of 'Aloha Aina'. It is a Hawai'ian expression that, literally translated, means 'love of the land'. For indigenous Hawai'ians like him, whose ancestors came to Hawai'i from other Pacific islands in tiny boats, 'Aloha Aina' is hugely important. 'It shapes our view of the world and the actions we take,' says Atay. 'Everything is connected, has an effect and is being affected, land, soil and life on it, air, water and the ocean. With our agricultural system, clean water came from the mountains through our fields and went back into the ocean. The cycle of life in the ocean provided food and everything else we needed. We didn't just survive here for thousands of years, we thrived.' Harmony, the connection between humans and nature, the interaction of earth and ocean, wind, water, and rain, is for Hawai'ians like Alika Atay as important today as it was for his ancestors. It comes with an inner certainty as to what it means to be Hawai'ian. 'And then these ag companies show up and tell us genetically engineered crops are better for us,' says Atay, shaking his head. And there are pragmatic insights: 'If the environment and the ocean die, there will be no tourism anymore.'

Alika Atay's decision to get involved in the anti-pesticide campaign was only partly the logical conclusion following from the 'Aloha Aina' concept; it was experiencing pesticide poisoning first hand that brought home the need for immediate action. For years, Atay was the canoe coach at the High School in Ma'alaaea Bay, north of Kihei. The traditional ocean-going outrigger canoes are made from a single tree trunk – the first Polynesians came to Hawai'i in boats like these. For centuries, the canoes were used for fishing, to transport people and goods between the islands, and they played an important role in various religious ceremonies. They were a means of transport, a symbol of status and power, and survival depended on them. In particular for young indigenous Hawai'ians, being a member of a good canoe team is extremely important; it's an expression of identity and self-worth. December and January are high season for outrigger canoes, 'We were training every day,' says Atay. One day, in December of 2006, a big storm swept across Maui. Enormous amounts of rain fell within a very short time, creeks and rivers burst their banks, and fields were flooded. Pesticide-laden run-off from agricultural land, including Monsanto fields, flowed straight into the ocean. When the students resumed their training in Ma'alaaea Bay, they had no idea how contaminated the water there was. 'The kids were coughing and wheezing on the water, they felt dizzy and got headaches. The bay was drenched in chemicals,' Atay recalls. He stopped the training, got the students off the water, and called the department of health. For the next ten days, officials tested the water in the bay morning and night. Contamination levels remained so high that swimming in the bay was banned and no one was allowed to go out in a boat. When the anti-pesticide campaign started, Atay tried to find records of the measurements taken by the health department at the time, but none could be found – all documents relating to the incident have disappeared.

When the agrichemical companies went to court in 2013 to get Bill 2491 overturned, it seemed as if the anti-pesticide campaigners had run out of options. Activists on Maui therefore chose a different route: a ballot initiative. US states give citizens the option to initiate new laws or changes to existing laws through a referendum. On Hawai'I, that first meant collecting signatures; as soon as enough voters have signed a petition for an initiative or legislative change, the measure has to be put on the ballot at the next congressional or senate election. The activists on Maui proposed two independent studies. One was to assess the impact of the high levels of pesticide use on the health of residents living on Maui and neighboring Molokai. The other study was to be an environmental impact assessment. Both studies would be paid for by the agrichemical companies and until the final reports were ready, no genetically modified crops were to be planted. 'It was the first time the citizens of Maui used the democratic process of signatures to get a bill on the ballot,' says Alika Atay, who has a huge smile on his face when he talks about the 2014 campaign. 'The practice of a daily use of RUPs (restricted use pesticides) is not in line with our Hawaiian values of Aloha Aina. The industry is taking decisions that are affecting my children's grandchildren. My pitch during the campaign was simple: you live here, it is your home, and if it is your home, you have to protect it, regardless of economic gains and jobs. It is not about the gains for today, it is about the future.' When it became clear that a moratorium for growing GE crops might be on the ballot for the 2014 Midterms in November, the agrichemical companies started a campaign to get the bill dismissed on election day. The 'chemical cartel' rented billboards and sent out leaflets, they bought huge amounts of time for campaign ads to be run on radio and on TV. 'You just could not escape the pro GMO messaging,' says Ashley Lukens from CFS. 'It was a sleazy, creepy advertising campaign that blanketed the state and people could sniff it.' Lukens and the Center for Food Safety did their best to provide activists like Alika Atay and Autumn Ness with material on pesticides and GE crops – leaflets, graphics, statistics, and research data. Autumn Ness gave up her job to lead the campaign on Maui. She recruited a small army of volunteers to knock on doors and speak to residents about the possibility of a ballot initiative. Unlike Monsanto, the anti-pesticide campaign had no money for glossy brochures. What the activists did have was time, enthusiasm and photocopies of a power point presentation prepared by the Center for Food Safety. 'We wanted to get a very simple message across: 'Tell us what you are spraying',' says Luken. 'There is no merit in going toe to toe with the industry on abstract science. Instead we needed to repeat that the vast majority of GE crops are pesticide tolerant and that the companies fight disclosure and oversight.' But even a simple message needs to be heard. Autumn Ness personally knocked on 3,000 doors, altogether the tiny army of volunteers managed to talk to 10,000 residents – there are only 140,000 people living on Maui. On Hawai'I, it is a legal requirement that organizations and companies engaged in a ballot campaign disclose their campaign spending a week before the election. The anti-pesticide campaign had spent $200,000, Monsanto, and the rest of the 'chemical cartel' had invested almost 9 million dollars. 'I just said to people: who will spend over $8 million to tell you the truth?' says Alika Atay. 'The spending report was a game changer,' says Ashley Lukens. 'We knew we could win on the strength of people's frustration and the fact that the ag companies spent millions to avoid regulation. People really hated that these companies had tried to manipulate the democratic process.' When the election results came in, it soon became clear that a majority of voters were in favor of the proposed moratorium – but as on Kauai, on Maui, too, the agrichemical companies immediately challenged the decision in court.

Even before the GMO moratorium initiative campaigners on Maui had tried to find out as much as they could about Monsanto and what the company was doing on Hawai'i. Some information is available through the USDA and EPA websites – if one knows what to look for and manages to find the relevant links. It is possible to assess on a monthly basis what quantities of restricted use pesticides have been used on each island, but the amounts cannot be linked to individual companies.

The USDA has to licence field trials with a newly developed genetically engineered crop. The companies have to submit applications, describing the scope of the planned tests. The authorities are supposed to consider them and either grant a licence or withhold it. From his time as senate majority leader, Gary Hooser knows where to go and whom to ask about permits requested by the agrichemical companies. At the Health Department, he found the applications submitted by the agrichemical companies – most of them in unopened boxes that were filling a whole storage room. The relevant departments had given their approval without even looking at the applications.

The activists on Maui began by trying to locate the Monsanto test sites. While a number of different agrichemical companies operate on Kauai, Maui is 'Monsanto territory'. The office complex is in Kihei, right next to Route 31. It's not sign-posted but easy to find anyway: take the exit for the police station, pass the car park and you are right outside the Monsanto gate. Activists say it is the only existing entrance to the Monsanto test sites. It took campaigners a whole day to follow the fence that surrounds not just the office buildings and greenhouses but all of Monsanto's agricultural land. In some sections, there is just a single fence, in others there are three rows of very high fences secured with razor wire. 'It's like the frigging Berlin Wall,' commented one activist who has seen these sections of the fence. Other test fields have a different kind of protection: they are situated in the middle of sugar cane plantations and can only be spotted from the air – if one knows what to look for. There are good reasons for Monsanto to shield the test sites as best as they can from the public in general and nosy activists in particular. One eye witness described to me that the test fields are divided into tiny plots[9]. A spray table lists the individual plots and when which pesticide is to be applied. Over weeks, the area will be sprayed with a variety of pesticides on a daily basis. Autumn Ness confirms the use of extreme amounts of chemicals on the test sites. While she was going door-to-door during the 2014 ballot campaign, she was able to talk to a number of field workers. 'We sprayed the shit out of that stuff, every few days we sprayed the same field,' one worker told her. Most of the Monsanto field workers come from the Philippines and are in the United States on a nine-month work visa for agricultural laborers. At Monsanto, safety instructions for pesticide use are given in English and Filipino. Several workers told Ness that they have to provide urine samples on a regular basis but are not informed about the results. While the Filipino workers assumed they were being tested for drugs, Autumn Ness

---

9   I have met and interviewed this eyewitness in person but am withholding witnesses' identity for their protection.

is convinced that they are tested for pesticide residue. Health professionals share that view.

Corn plants in the Monsanto greenhouse on Maui

Since 2000, Dr Lorrin Pang has worked for the Hawai'i Department of Health. He is an endocrinologist, and before he settled on Maui, he cared for veterans at Walter Reid Hospital. For 25 years, he was a consultant to the World Health Organization, and he is still consulting for the humanitarian non-governmental organization Doctors without Borders. Because of the international recognition of his work, his knowledge, and his long experience, his bosses would dearly love to get rid of him. Since he was an expert witness for the residents on Kauai who sued DuPont/Pioneer over the continued exposure to pesticide-laden dust, he is not allowed to talk to the press in his capacity as senior staff member at the Hawai'i Department of Health, nor is he allowed to speak for the department. 'Everything I am telling you is my private opinion as physician and scientist,' is the first thing Dr Pang mentions when we meet in a small café in Paia. The DuPont/Pioneer court trial was extremely important, says Pang. 'Under disclosure Pioneer had to show which chemicals they used. That's how it became known that this company used 85 different chemicals, pesticides, and enhancers. There are trillions of possible combinations of these chemicals, and even if each one of them was safe on its own the additive and cumulative effects have not been studied. Nor has the damage overlap been studied: for example, if one chemical damages the liver, what damage do other pesticides do to that already weakened organism?' For scientists like Lorrin Pang, it is almost impossible to study the effects that pesticides have on human health. He tells me about a group of Filipino workers who had to fill seed into sacks marked as 'not for human consumption'. Normally, this precaution would be taken for coated seeds: to make sure that the seeds farmers drill germinate and are not attacked by pests, they are given a protective pesticide coating. The workers had various respiratory problems and Dr Pang was worried that the symptoms might have been caused by pesticide dust. It would have been easy to find out whether that was indeed the case; all he would need to do was a medical exam and a few tests. But the workers were scared they would lose their jobs if they talked to Dr Pang and refused to be assessed. The only way Dr Pang was able to communicate with the workers was through a legal assistant who passed questions and answers back and forth.

So, the million dollar question is: what are the agrichemical companies doing on Hawai'i? Why are the fields protected by tall fences? Why the secrecy, the warning signs that trespassers will be prosecuted? Why do these companies go to court as soon as they face the prospect of having to disclose what they spray? What crop or procedure could possibly require such extraordinary amounts of pesticides? These are questions I wanted to put to representatives of the agrichemical companies on Hawai'i. Several weeks before we were scheduled to travel I started putting in interview requests – a task that soon turned into what is every journalist's nightmare. At Syngenta I was told to contact Hartung Brothers, Inc., because Syngenta no longer had a presence on Hawai'i. Hartung Brothers, Inc. is based in Madison, Wisconsin – that's about all the information the company provides. The website is almost impenetrable, there are no phone numbers to be found, and the only way to contact the company is via an online request form – which in my case was being ignored. Interview requests have to be sent to the 'Hawaii Crop Improvement Association', HCIA, I'm told by another company. HCIA is the interest group representing all agrichemical companies on Hawai'i and solely responsible for dealing with media requests. 'WE'VE GOT ANSWERS. We're eager to help you understand crop improvement, crop protection, ag biotech, GMO safety and our role in Hawaii agriculture. Fill out our form and we'll get back to you soon! Mahalo!' the website informs me. And HCIA do get back to me – unfortunately they are unable to deal with journalists; for interview requests I am to contact the PR company Becker Communications Honolulu. At Becker Communications, I am the one who has to answer questions. For whom will I be writing? Will the text have a pro or anti GMO stance? My answers seem to be unsatisfactory, my emails remain unanswered, phone messages are not passed on, the members of the PR team who could deal with my request are either in a meeting, out to lunch or on holiday. Then someone recommends to try Monsanto directly – the website offers visitor tours upon request. But here, too, my calls remain unanswered. Then I get lucky: when I once again call Monsanto in Honolulu, a security guard answers, the building has just been evacuated because of a fire alarm. I explain that I am calling from London and because of the eleven-hour time difference, it is already late evening here, so can he help me out with a direct number. And so, an hour later, I am talking to an entomologist at Monsanto. He can't help me book a tour, but when I keep emailing him the next day he does pass on my request to a member of staff in the Monsanto outreach team. A month later, we are on Route 31 on Maui, take a left for the police station in Kiehi, pass the car park and stop in front of the Monsanto gate.

That morning, no fewer than three Monsanto staff have flown from Honolulu to Maui to meet us. We are sitting in a large conference room in an otherwise empty building – because of the storm and the blackout, the offices remain closed today. Research into genetic engineering is done in the laboratories at the Monsanto headquarters in Missouri, explains Dr Michelle Starke, responsible for science & environmental outreach at Monsanto. The work on Maui is all about breeding perfect parent lines for the more than one thousand hybrid seed varieties Monsanto sells mainly in North and South America. The farmers need seeds that are perfectly adapted to the climatic and weather conditions in their growing region. In northern states, winters usually are long and the growing season short, which doesn't leave much time for corn and soy to grow and come to maturity. In southern states, the growing period is much longer, leaving crops more time. East of the Mississippi there is usually plenty of rain, but the West is dry, and corn and soy either have to be irrigated or cope with a lot less water.

The agricultural land next to the office complex lies fallow; on some beds, cover crops are growing. We don yellow safety vests with reflective stripes, get on a company bus and drive a hundred yards to the nearest greenhouse. Inside are thousands of corn plants in different stages of development. Each has its own label with a barcode that makes it clearly identifiable. The female inflorescences are covered, the males are stuck in bags so that the pollen they are shedding will be caught. Pollination is done by hand and the scientists know and control the heritage on both the male and the female side. In field trials, plants then have to prove whether they are suitable to be part of a parent line. On neighboring Molokai, large volumes of all parent plants are bred. The seed is then sent out to contract farms on the US mainland, where breeders produce the hybrid seed that is then sold to farmers, explains Dr Starke. And of course, small amounts of pesticides are used in the production of parent lines. Our visit was scheduled to last an hour, we've been talking for three. The four staff members remain friendly, non-committal and vague. Critical questions stay unanswered, and follow-up questions yield no results.

It is not true that Monsanto and other agrichemical companies only breed parent lines on Hawai'i. Experimental field trials have to be licensed by the USDA, and the applications by different companies are listed on a publicly accessible USDA website. The testing of genetically engineered crops began in 1987. Since then, more than 3,000 licenses have been approved for Hawai'i, more than for any other state[10]. Ninety percent of applications are for genetically engineered corn and soy, the other ten percent are for cotton, alfalfa, canola, and sugar beets. Genetic engineering focusses almost exclusively on two traits: resistance to herbicides and to pesticides. The companies 'routinely withhold key facts about their GE crop field tests as "confidential business information" (CBI). In fact, 81% of Hawai'i permits (in 2014) hid the identity of one or more traits as CBI,'[11] finds the study by the Center for Food Safety.

But why and for what are such enormous amounts of pesticides needed? In the US, corn and soy are grown on roughly 180 million acres (73 million hectares), about 90 percent are GE varieties. 'Corn accounted for 50% of all herbicides (by weight) applied to 21 major and minor crops in the U.S. in 2008 (USDA ERS 2014), including about 85% of the endocrine-disrupting herbicide atrazine,'[12] says the CFS study. Ninety percent of all corn seed is treated with neonicotinoids.

In seed production, a significantly higher amount of pesticides per unit of area is needed. It is rare for farmers to save their own seed for the next season. They need high-yielding varieties, year after year, and the seed should meet the same standards for quality and homogeneity. Only hybrid seeds are likely to meet such criteria. They are relatively robust, and there is a high probability that each plant will perfectly express all the desired traits.

To produce hybrid seeds, genetically different plants are chosen and crossed. In corn, one parent plant may hold the genetic traits to produce large cobs with big kernels. Another parent plant may not grow very tall but come to maturity in a very short time. If these parent plants are crossed, the resulting offsprings should deliver a high yield, even with a late sowing date. But this generation of

---

10 Center for Food Safety: Pesticides in Paradise, 2015, page 22ff
11 Lit.cit. page 25
12 Lit. cit. Pesticides in Paradise page 29

hybrid plants usually is not 'true to seed'; there is only a small probability that the next generation will express the desired traits. Instead, different and new combinations of genes are likely. Therefore, farmers hoping for high yields will have to buy fresh hybrid seeds every year.

What agrichemical companies like Monsanto produce on Hawai'i are the parent lines needed for mass propagation of seeds. The one way to make sure that a plant is highly likely to pass the desired traits on to the next generation is inbreeding – not for one but over several generations, the plant is crossed with itself. The higher the degree of inbreeding in both parent lines, the higher the probability that the traits the next generation will express can be accurately predicted, says Dr Ricarda Steinbrecher, molecular geneticist at the British non-governmental organization EcoNexus. The price for the extreme expression of one trait is that others are underdeveloped or missing. A plant might only develop a shallow and weak root system. As a result, it will be malnourished and require large amounts of chemical fertilizer to make up for it. Such plants will be more susceptible to diseases and likely to prove a magnet for different pests. In addition, the roots will lose most of their ability to communicate with the mycorrhiza, says Steinbrecher. That exacerbates the malnutrition, which has to be compensated for using more chemical fertilizers. And chemical fertilizer is for plants what fast food is for humans: to live on a fast food diet long-term will make humans and plants ill. And on top of all this, these weak plants have to cope with genetic modification. The endocrinologist Lorrin Pang and Hector Valenzuela, professor and vegetable crops extension specialist at the University of Hawai'i at Manoa, work in very different fields, but they both understand genetics and come to the same conclusion why the pesticide need for GE parent plants is so huge. Genetically modified plants show a defense action like humans will after an organ transplant. The plant will use all its resources to fix that problem and has none left to deal with fungal attacks, pests, or other diseases. To protect the GE plants and increase their chance of survival, they are grown in an almost sterile environment. And this environment is created by using large amounts of pesticides – the spraying schedule the activists found on one of the Monsanto fields suddenly makes sense. Pesticides are sprayed to prepare the seed bed, before the seeds are in the ground. The soil will be fumigated: treatments with pesticides, fungicides and herbicides ensure that there are no pathogens or weed seeds left that might harm the GE plants. In the process, the soil biology, microorganisms, earthworms, and the mycorrhiza network are obliterated. What remains is sterile dirt and pesticide-laden dust which is blown with the wind across the residential areas nearby.

Another factor that is a massive problem for growing GE corn is the tropical climate on Hawai'i. It is much too hot and humid for growing corn, and that makes the plants even more vulnerable. Nevertheless, the process is worth it for the agrichemical companies. It takes about five successive generations of plants until it is possible to assess whether a particular trait – for example high yield – is truly established in the plants and will be passed on. Only once that's confirmed is a plant suitable for use as a parent plant. In Hawai'i's tropical climate, three crops per year can be grown and harvested. In the Midwest, growing five generations of plants and finding out whether they are suitable parent plants would take five years – in Hawai'I, it can be done in a year and a half. But the work isn't finished once the 'optimal point of inbreeding' in a plant has been reached. The whole process has to be repeated immediately: the optimised parent lines will do their job for a year, but they are unlikely to remain stable after that. Parent lines have to be maintained, corrected, adapted, and optimised on a continuing basis[13]. The same goes for traits created through genetic engineering (like glyphosate tolerance) – whether they will definitely be passed on to the next generation is not a given. That is why GE plants have to be repeatedly sprayed with the relevant herbicides throughout the growing season.

The production of genetically modified conventional hybrid seeds is at the core of what agrichemical companies do on Hawai'i. GE technology provides the all in one farming toolbox: each seed variety is sold with the right kind of chemical fertilizer and pesticides. Farmers don't buy just seeds, they buy a 'crop system'. The work on Hawai'i is essential for companies like Monsanto, and is a fantastic way of maximizing profits. Not only does Hawai'i have the perfect climate, but labor costs are low, there are few environmental, health, and safety requirements and even less oversight: the constraints that do exist are often not enforced. Hence it is not surprising that the 'chemical cartel' immediately goes straight to court when – first on Kauai, then on Maui and finally on the Big Island – bills are passed that threaten to limit what companies can do[14].

---

13 The reason for that is 'epigenetics', says Ricarda Steinbrecher. Epigenetics deals with the activity and function of genes. An organism can mute or activate particular gene sequences. The gene may be passed on in its entirety to the next generation, but the organism will 'decide' whether to reactivate or deactivate genetic sequences depending on whether their functions have been proven to be useful or superfluous.

14 On Hawai'i's Big Island, a bill banning GE crops was passed as a precautionary measure, the residents feared that the agrichemical companies might simply move their operations from Maui and Kauai to the Big Island.

George Krimbrell is a lawyer and legal director at the Center for Food Safety. Legal fights over GMOs and pesticide use resemble those between David and the agrichemical industry Goliath. George Kimbrell has been in legal battles against the ag industry more than once and, like the biblical David, he's been pretty successful. We sit in his office in Portland, Oregon. Outside it's raining, as it would on an October day in the Pacific Northwest. The court on Hawai'i merged the cases the agrichemical companies brought on Kaui, Maui, and the Big Island, says Kimbrell, who, together with Ashley Lukens, the CFS director in Honolulu, was involved in the dispute from the very beginning. 'Our role was to make the laws as strong as we could because we knew the industry would challenge them. We were the national group that was most involved,' he explains. The 'chemical cartel' won the first round. The district judge decided that the bills were pre-empted by federal law. The federal government takes decisions in regard to GMOs, and pesticides and federal law trumps state law. The cases went to the 9th circuit court of appeal, the second highest court in the US, which covers Oregon, Washington, California, and Hawai'i. The court ruling came in the fall of 2016, and it had good and bad parts. The judges decided that counties are explicitly pre-empted from regulating matters having to do with pesticides or genetically engineered seeds, but states can. And so can counties, if the state grants them such powers. Legally, the ruling opened a door: 'The take home is: the state of Hawai'i could enact a law that protects its citizens and the environment from the harm of pesticide spraying and the accompanying genetically engineered crop planting, and the federal government can't stop them.' The state government in Honolulu could rule on buffer zones, it could force the industry to disclose which pesticides they spray and when, it could decide on a moratorium for planting GE crops and enable the respective counties to pass bills that take the particular circumstances of each of the islands into account. The activists who queued overnight on the pavement outside the county council building on Kauai, the volunteers who went from door to door on Maui, the campaigners who organized demonstrations and distributed leaflets – are they the ones who now have to change the minds of legislators in Honolulu? And what about the paid lobbyists working for the agrichemical industry warning of massive job losses and waning tax returns? George Krimbrell remains optimistic, at least in the long run: 'Do I think that's going to happen any time soon? No, I don't. What do I do? If you are a hammer, everything looks like a nail. I am a lawyer, I use the law the best way I can to stop the bleeding, that's what our cases are about. And also, because legal cases can create societal change. The way people think about an issue changes because of the cases. I'll give you an example. He points to a framed photo on his office wall. It shows Kimbrell and his colleagues on the steps of the US Supreme Court in Washington D.C.. 'That is the only case I ever had that went to the US Supreme Court. That's the case against Monsanto in 2010 about genetically engineered alfalfa. It took three years of my life, but it was a unique experience, and, more importantly, that case changed the way people thought about genetically engineered organisms because there we had a legal decision on cross-pollination, on contamination of organic and conventional farms.' Until then, Monsanto had argued that cross pollination through bees or wind wasn't an issue because except for the herbicide tolerance, GM alfalfa, corn, or soy were exactly the same as conventional or organic alfalfa, corn, or soy. And because Monsanto seeds were herbicide-resistant the product actually was superior to non-GE seeds, Monsanto GE alfalfa in a field of conventional or organic alfalfa therefore was a 'favorable presence' and not 'contamination'. Hence, Monsanto was not obliged to contain it and prevent cross pollination.

'We said: no. You're losing diversity,' says Kimbrell. 'You're losing the farmers' ability to choose and the consumers' ability to choose what they buy by allowing contamination like that.

'The Supreme Court said: yes, you are right, that is a recognisable legal harm. Since that time, contamination matters. It matters legally, you are seeing all the class action cases now, suing for damages from contamination, etc., it matters in the terminology, the way we talk about it. Nobody calls it 'advantageous presence' anymore. Now people call it contamination.'', says Kimbrell.

I had met George Kimbrell for the first time in 2005 at a conference for organic agriculture. It wasn't just his eloquence and the impeccable logic of his arguments that fascinated his audience, but his passion. For him, it is about far more than a victory in court: 'You've got to shift the consciousness of the populous, you've got to awaken people to the realities and show them that there is an alternative to Monsanto's vision of our food system, a paradigm shift. But most of the litigation is tourniquets on all the catastrophes that are happening. And the way we do that is, we look for existing legal mechanisms. The problem here in the United States is that we have a dysfunctional legislative process, and so most of the laws I use are 40 to 50 years old. They were never intended to legislate genetically modified organisms or modern pesticide issues. Basically, you are squeezing blood from statutory stones. You are trying to use square pegs in round holes. And you do the best you can. Those are the levers until we have a new age of progressive legislation. There is not a framework yet in place that would be similar to the precautionary principle that you see in European law, there is some of it in American laws, but not as much yet. We focus on the environmental side of things because we have very strong environmental protection laws. But it's trench warfare, it is literally trench by trench warfare. Two steps forward, one step back, like in Hawaii, we've lost some, we've won some and that's the way it goes because we are fighting against probably the most powerful industry in America, the chemical industry.'

On Kauai Gary Hooser uses a passage from the state's constitution – the state in all its political subdivisions shall protect waters and the environment – as leverage to initiate environmental reviews. The courts will now have to decide whether an environmental assessment should have been done when Syngenta sold its holdings on Hawai'i to Hartung Brother's, Inc.. Both Nene geese and monk seals are protected under the endangered species act. With Syngenta renting some public land, should their rights not have been protected through an environmental impact study, a process that requires the companies to disclose all the pesticides they use?

Hooser is also suing government departments which gave permits to agrichemical companies to plant crops that have not been deregulated without properly evaluating the applications. The required environmental impact statements were not done, and the Department of Land and Natural Resources has not reviewed these documents, says Hooser. They did not even open the boxes.

And Hooser and his fellow campaigners are taking on board the lessons learnt from the decision of the 9th Circuit Court: a county may not have the right to regulate the pesticide use of an agrichemical company or insist on buffer zones. But a county has the power over land use. On Big Island a bill will define what constitutes a large animal operation. Big ag operations close to residential areas or an environmentally sensitive area would need a permit which will only be given after a public hearing and can come with conditions attached. If the bill were to pass it could become the blueprint for similar bills on Kauai and Maui.

These are just some example for the legal steps campaigners on Hawai'i have taken to make the activities of the agrichemical companies more transparent and regulate the pesticide use on the islands. For lay people, the details may seem very complex and somewhat confusing, but Gary Hooser and his colleagues are well on their way to getting a square plug into a round hole.

In the US, environmental law still offers the best leverage to effect change. Pesticides can cause health problems – that is not disputed: pesticides are poisons. What is extremely difficult to do is proving that a specific pesticide has caused or contributed to a health problem of one or more individuals. Were they even a factor? For a physician like Lorrin Pang, it's the wrong question to ask. Because of the pesticide use on Hawai'i, in his opinion, all its residents have become unwitting participants in an 'illegal experiment' to which they never consented. He compares the situation on Hawai'i to the US soldiers in the 1960s who were exposed to high levels of radiation during nuclear tests conducted by the US military. A 1994 senate hearing ended with the conclusion that exposing people without consent is akin to human experimentation. 'In regard to pesticides, the same is true on Hawai'i,' says Lorrin Pang. 'My issue is not with the agrichemical companies, it is with the regulators.'

CFT's legal director George Kimbrell agrees with Dr Pang that the pesticide use on Hawai'i does amount to a big experiment, but in legal terms that does not mean much. The Center of Food Safety takes a different approach: it's about consumer choice. Consumers have a right to choose between conventionally produced fruit and vegetables or certified organic produce, they have a right to buy eggs from caged hens or choose free range ones, but only if they know whether a food product contains GMOs can they exercise their right to exclude them from their diet. Organic produce will be certified; on any egg carton the consumer will find information about the conditions under which the laying hens are kept, but so far US customers have no way of knowing whether a product contains GMOs or not[15].

Food companies and agrichemical companies spent millions of dollars in an attempt to squash the different GMO labelling initiatives. Measures for voluntary labelling are in a consultation phase. One idea is to have a QR code on the product so that consumers can access the relevant information via smart phone. And the industry wants to get away from the terms GMO and genetically engineered and would like to talk about 'bio-engineered' products instead. But consumers remain wary and mistrust the industry. Many food producers have recognized that and now clearly label products as GMO free.

Consumer mistrust is justified, says George Kimbrell. 'What we tell people at the Center for Food Safety is: the one thing you want to know is the US Food and Drug Administration[16] doesn't do any independent testing of the food safety of these foods. They have a voluntary meeting with Monsanto or Syngenta or Bayer or whoever it is. That entity shows them the summary of their health studies, not even the real studies, the summaries of them, and they say they have no questions. That happens in a confidential meeting, I can't attend it, and even that meeting is

---

15 Organic products are the exception because the use of genetically modified ingredients is generally not allowed in organic production.
16 The Food and Drug Adminitration is a federal agency of the United States Department of Health and Human Services. The FDA authorizes the use of new drugs and supervises food safety standards.

voluntary. And that's the extent of our food safety assessment and "approval" in the United States. We don't really have a food safety approval process for human health risks of genetically engineered foods. It is an experiment to the extent that we are taking these companies, their research and their word for their safety, to the extent that there are no long-term studies being done, and that there are no independent studies by universities being done. All of these things are huge red flags. And the past is prologue. So many other companies have told us: don't worry this is totally safe.

'I don't think we have smoking guns that genetically engineered foods are killing anyone tomorrow. But I also don't think we have evidence of their safety. And it is an on-going experiment as to what the long-term effects are on us who are eating them.'

The environment, climate change, agriculture, pesticides, health, and food – to George Kimbrell, it doesn't make sense to look at any of these issues in isolation because they are interconnected. 'How we eat and how we feed our families is probably the most direct and intimate way we interact with the environment every day,' he says. I consider myself an environmental lawyer, that's my training, and I do food law. And that's because food and agriculture is just a new branch on the tree of environmental law. It is the future of that world.'

Gary Hooser, Autumn Ness, MiKey Boudreaux, Marghee Maupin, Ashley Lukenss, Alika Atay – they all share this view, but after the long fight to regulate pesticide use on Hawai'i, the dashed hopes, and the disappointment over the court decisions, a certain hopelessness seems to have spread, few have the energy to think about what might come next.

Gary Hooser is working on a number of legislative initiatives, and, together with H.A.P.A, the Hawai'i Alliance for Progressive Action, he is trying to motivate young people to stand for public office. Alika Atay is one of several Hawai'ians of Polynesian decent, who has campaigned for the anti-pesticide bill on Maui, and who has realized that with time, energy, and enthusiasm, the activists are able to have an impact, even on a company as powerful as Monsanto. And this new found confidence remains: in 2016 Atay stood in the county council elections and was elected. And there is one other thing that has changed: 'We made so many friends during the campaign. When we won, a white guy came up to me, slapped me on the back and said: "Who would have thunk the 3 Hs would do it: Hippies, Haoles and Hawaiians!' The hippies live the values of Aloha Aina every day. Haoles choose to live here, centred around organic food, healthy life style, education for their kids, and the Hawaiians were raised with Aloha Aina.' And the

Filipinos have been part of the fight as well, says Alika Atay. They understand the concept of 'nature' and living with nature. Hawai'i has been shaped by the brutality of colonialism and plantation agriculture as well as industrial exploitation. It is an island state in which many ethnicities have to live side by side and different cultures, and traditions and social norms clash on a daily basis. With that in mind, it is hard to overestimate the importance of 'Aloha Aina' suddenly becoming a leitmotiv not just for indigenous Hawai'ians. And Alika Atay's assessment wasn't just valid for the short-lived euphoria after the successful ballot initiative. A few days earlier, in rainy and cold Portland, George Kimbrell, too, got a little emotional talking about Hawai'i:

'You know the cool thing with Hawaii is that… it's been really remarkable to do the work there and help the people because it's a spiritual work. The people there have a spiritual connection to their food that is very unique in America. In Hawaii, because they have this history of colonialism, that's so powerful – being able to work with the native leaders there, like Walter Ritte and other native Hawaiians that have fought for Hawai'ian independence for 30 or 40 years – they are real heroes. And they are real warriors, spiritual warriors. To have them connected to our movement about food and about the environment – I think for us to succeed as a movement globally and nationally, we have to have that, everywhere.

'To see that there is a seed of that in Hawaii, the way people think about taro[17] and the islands as their home and these companies as colonialists that they need to kick out and choose differently for themselves and have independence for their own agriculture and not import 70 percent of their food like they do at present, that's a powerful thing and a hope. I loved working there'.

And in spring of 2018, the activists' continuous hard work on Hawai'i paid off: on April 6th, the House of Representative in Honolulu unanimously passed SB30095. The bill requires agrichemical companies on Hawai'i to disclose which pesticides they use. Schools get a 30-yard buffer zone, restricted-use pesticides cannot be sprayed during school hours. And the use of the insecticide chlorpyrifos will be phased out – chlorpyrifos is an organophosphate known to harm the nervous system in humans. A few days after the vote, the activists organized yet another demonstration to get the Hawai'ian senate to approve the measure, too. 'Wear red and show that you are with us' was the slogan for the protest that was supported by 20 different organizations. The group photo taken outside the parliament building in Honolulu shows Hawai'ians next to Hippies and Haoles – and a sign

---

17 Taro is a tropical food plant with a very starchy root. On Hawai'i is a traditional staple food crop and as important as potatoes are in many European countries.

stating 'We are Aloha Aina'. On May 1st, the state senate passed the bill, and Hawai'i became the first US state to phase out and ban the use of chlorpyrifos.

The agrichemical companies on Hawai'i are definitely beating a retreat, campaigns and legal challenges have cost millions, Hawai'i has become just too expensive. For now, this only means that the industry is moving elsewhere. Puerto Rico is becoming popular. Autumn Ness closely monitors the relevant USDA websites and is seeing evidence of such a move. Field trials have to be licensed, and Ness shows me the website: 'The location always used to be HI for Hawai'i, now it's mostly PR for Puerto Rico.'

'The Center for Food Safety has already contacted campaigners in Puerto Rico, the experience we gathered during the campaigns in Hawai'i will be really useful,' says Ness. 'We will help them in every way we can.' We leave unsaid what probably is on both our minds: just four weeks have passed since hurricane Maria devastated Puerto Rico with 170 mph (nearly 280kmh) wind speeds. There is still not enough food and clean drinking water, and no one knows when the electricity supply will be fixed. Reconstruction will take years, and Puerto Ricans will have little time to fight off agrichemical companies.

George Kimbrell remains optimistic. He is convinced that in the long run, GMOs have no future. 'I think it's a zombie paradigm. You are going back 25 years now, or even 30, 40 years to the research in the 80s, there is the hype and the myth of stopping climate change, of feeding the world, of increasing yields, none of which is true,' he says. 'There is just one thing they became really good at, which is engineering crops with a resistance to their own pesticides. And that's it, so it's really kind of a one trick pony. I don't see any evidence from our scientists or from other sources that that's changing and if it doesn't, then the question is: where does American agriculture go next as far as the future of the food is concerned, but I don't think it's going to be genetically engineered organisms. I think right now the problem is: zombies can cause a lot of damage, and that's what's happening.

'I think genetically engineered crops have been such a controversy, rightly so, because they are a microcosm of this larger debate over industrial agriculture. The pesticides, the kind of crop system they created, it's one of the major pillars of how we produce our food right now in America. It supports the animal factory system. Without the feed you wouldn't have these confined animal feeding operations, the way they are set up and all the harm, the tragic immorality, that they have caused. How do we solve this? I don't think there is a silver bullet.

'I think we are at a very critical moment, it's kind of on a knife's edge as to what happens next. I'm optimistic that there is a good deal of hope just because I see the paradigm, the technology is fundamentally flawed and limited in a way that makes it kind of a zombie walking. But how fast that happens and what harm it causes in the interim, those are the real questions.'

Monsanto test site on Maui

For the past few years Bayer-Monsanto, has been swamped by law suits. In June of 2020, the company reached a deal with claimants, the company will pay out 10 billion US$ to resolve roughly 75 percent of the claims. An estimated 125,000 people allege that exposure to glyphosate caused them to develop cancer. The settlement excludes the trial cases in which the plaintiffs were awarded more than 2.3 billion US$; Bayer-Monsanto has appealed against the verdicts. The company continues to deny any link between glyphosate and non-Hodgkin lymphoma. Roundup products remain on sale in the US and will not have to carry a warning label. With this settlement, Bayer-Monsanto manages to contain the damage somewhat, but the company's legal troubles are far from over.

In 2019, the direction the zombies are taking is becoming a lot clearer. Many thousands of cancer sufferers are suing Bayer-Monsanto; they got ill because they were exposed to glyphosate, the active component in Roundup, Monsanto's best-selling herbicide. Their claim that Monsanto knew about the potentially carcinogenic effect of glyphosate but didn't warn users was backed up in court: internal emails and communications between Monsanto and the EPA told a story of lies, deceit, coercion, and contempt. In several cases, the victims were awarded huge sums in compensation, but Bayer-Monsanto went to appeal and seem to be inching towards an overall settlement. A glyphosate ban seems to be possible, at least in Europe. Agrichemical companies may be looking back at the 1990s and the fate of the tobacco industry: as a consequence of the class action law suits, smoking outside the home has mostly been banned in Europe as well as in the US, and advertising tobacco products is very restricted.

In July of 2019, Gary Hooser took stock and reflected on how far the anti-pesticide campaign on Hawai'i had come. 'The USDA just reported that GMO seed corn production has dropped statewide by over 50%,' he wrote in his July 1st newsletter. 'Accompanying this reduction will be a parallel reduction in both general use and restricted use pesticides.' And then he quotes the latest USDA figures: in 2011/2012, the value of Hawai'i seed corn was US$ 250.000, and by 2018/2019, the value was down to US$ 106.000. In the same period, the total acreage for Hawai'i seed corn decreased from 6,910 acres (2,800 ha) to 2,530 acres (just over 1,000 ha). It is unlikely that these changes would have happened without the anti-pesticide campaign and the activists who fought so hard for buffer zones around schools and residential areas. 'With the increased public spotlight on their operations the increased legal actions in court, and the increased regulation at the State level, all combined have to have an impact on their business decisions as to where to base and grow their operations. And by now, these companies know that there is more public action to come, more negative media attention, more legislation and more actions via more lawsuits,' says Gary Hooser, ending with: #neverquit which is as much a statement as an appeal – the fight against agrichemical companies and the use of pesticides will continue.

Farming with Benefits

Ahupua'a – Hawaiians relearn growing their own food

Taro field

It's easy to miss the turn off for the Ho'okua'āina farm, Dean Wilhelm had warned us. No sign marks the narrow dirt track that rather suddenly branches off from the main road. After a short stretch, we are surrounded by tall trees; the taro farm is situated in the middle of a forest. Just one mountain range separates us from Hawai'i's capital, Honolulu. As the crow flies, it's a mere 10 miles (15km) to Waikiki, with its famous beach, luxury apartments, and high-end shopping. It is still early in the morning, but already the air is hot and humid. Except for the loud, unfamiliar calls of tropical birds, everything is quiet. To our left, the ground drops steeply, and through the trees we glimpse the taro field. The plants, which can reach over 6 feet (2m) in height, are planted in waist-deep water. As they grow, new shoots emerge from the fleshy stem and unfurl into huge, heart-shaped leaves. Narrow dams divide this section into 19 fields, each with taro in a different stage of development. The soil is almost black, which makes the taro leaves stand out in a vivid, nearly fluorescent green, whereas the dark water reflects the blue of the sky overhead, dotted with white clouds. Observing the seemingly still surface for a while, one begins to notice a very

Farming with Benefits

slight current and a myriad of tiny fish and tadpoles flitting about in the depths of the water.

Dean Wilhelm started planting taro ten years ago. He wanted to find out in practice how the traditional Hawai'ian growing system works, and whether it might even be possible to re-establish it. Taro – or kalo, as the Hawai'ians call it – isn't just a food crop but is deeply connected to ancient traditions, a way of life, and Hawai'ian identity.

A traditional Ahupua'a system would have reached from the accessible parts of the mountains right down to the coast where the water from the taro field would have fed into fish ponds.

Hawai'i is extremely fertile. If one 'farms with nature' and knows how to use the opportunities that climate and soil offer, it does not take much to produce plenty of food. The trade winds bring clouds heavy with rain which is released once they hit the high mountain ranges on the islands. The heavy rains create ideal conditions for tropical rainforests. Dense crowns of large leaves break the velocity and impact of the raindrops, and the humus layer which has developed over centuries can absorb and store the water. Slowly, it percolates though different soil layers until it finally reaches the ocean, where the cycle begins again with trade winds bringing rain clouds to the islands. The traditional Hawai'ian agricultural system is called Ahupua'a, and combines agroforestry with fish ponds and fishing. In the introduction and overview, a study[1] by the Center for Food Safety describes how the system worked: 'Hawaiians continued building their local food system by developing the ahupua'a land management system, which incorporated key principles of ecology and environmental management. The *ahupua'a* system organized the use of natural resources and access around self-sustaining land divisions that follow watershed contours from the tips of the mountains (mauka) to the near-shore fisheries *(makai)*. These wedge-shaped pieces of land encompass a range of microclimates and natural habitats, with each section uniquely suited to cultivating crops, hunting, fishing, and foraging. This arrangement allowed for maximizing the use of biodiversity over short distances and acknowledged the interactive influences of biological resources and production zones. This land management style preserved ecosystem services and provided advanced agricultural options that worked with the natural world. (…) An ahupua'a contained all the resources a community would need to sustain itself, while allowing for high productivity; historically it is reported that this system sustained a population of one million.'

The first settlers who came to Hawai'i brought chickens and pigs. The forests provided them with nuts, breadfruit, coconuts, citrus fruits, lychees, and edible ferns. There was wood for building boats and making utensils, while several plants provided different fibers from which cloths, sails, ropes and nets could be made – everything the settlers needed to go fishing. What forests and the ocean didn't provide was a sufficient supply of carbohydrates, and that is why growing taro is at the center of the traditional ahupua'a system, says Dean Wilhelm. On a flat piece of land, the Hawai'ians constructed a system of dams that temporarily caught the flow of water coming from the mountain. It slowly moved through the taro fields to be released at the lowest point and then made its way to the ocean. The starchy taro roots were used to make poi. 'You steam the roots until the skin comes off easily. Then they need to be ground, either on a poi board or with a pestle and mortar. The result is a kind of paste, not unlike potato mash but not as fluffy,' says Wilhelm, who makes poi every two weeks. His father was Swiss and came to Hawai'i to take a job as a chef at one of the hotels. He fell in love with a young Hawai'ian woman, they got married, started a family and decided to permanently settle on Oahu. Dean Wilhelm initially trained as a social worker, caring for prisoners and students from underprivileged families. He started to

---

1 Hawai'i Center for Food Safety: Pesticides in Paradise, May 2015, p.9 f
   https://www.centerforfoodsafety.org/files/pesticidereportfull_86476.pdf

notice how unsettled and rootless young adults and even children seemed to be. Wilhelm's own upbringing was shaped by the Polynesian-Hawai'ian traditions his mother had introduced him to and, like her, 'being rooted' is something he still takes quite literally. 'We need place-based education, connecting is the basis to find out who you are and define relations with others. Loss of the connection to food means loss of connection to place. If I care for the land, the land will care for me,' says Wilhelm. To Hawai'ians, taro was always central. It was one of the few food plants Polynesian sailors brought with them when they came to Hawai'i well over a thousand years ago. Taro is part of the Hawai'ian myth of origin, it is revered as the 'older brother'. To Hawai'ians, the kalo is literally part of their genealogy as well as the stuff of life'[2], writes the native Hawai'ian attorney and international indigenous legal expert Mililani B. Trask. Dean Wilhelm grows taro for sale, but visitors are welcome. And later that morning a group of school kids and their teachers will come for an introduction. Several benches are sheltered by a canopy. Next to it is a faucet. The freshly harvested taro roots have to be washed before they can be weighed and packed. With the help of a huge chart, Dean Wilhelm explains the growing cycle of taro. The kids listen attentively, but getting out into the field is even more fun: they are allowed to help prepare one of the fields for new seedlings. First, any remaining plant matter, weeds and old taro plants need to be removed, then everyone gets barefoot to trample down the seed bed until it is covered by a few inches of water. In the end, all the kids will be covered in mud and very happy, says Wilhelm. Before the fun begins, he demonstrates how he collects the taro roots from a plant that is about a year old and ready to be harvested. The parent plant is surrounded by new shoots. First, the leaves are cut off. They can be cooked and taste a bit like spinach. Then the rest of the plant is dug up, and the roots are cut off, cleaned, and processed. A single plant yields between three and five kilograms of root mass.

I get my first taste of taro at the Roots Café in Kalihi Valley, a suburb on the westside of Honolulu. Taro pie is made of shortcrust pastry with a lilac colored, slightly sweet topping that has the consistency of mashed potatoes – maybe taro is an acquired taste.

The Roots Café is part of KKV, a community and health center. Neighbors meet over coffee and cake for a chat; for KKV staff, the café serves as a canteen and meeting room, and everyone is welcome to just drop in for some shopping. There is a selection of fresh fruit, vegetables, dairy products, and bread on offer, all of which is much needed in Kalihi. It's only about five miles to Waikiki beach

---

2   Mililani B. Trask in Facing Hawai'i's Future, Hawai'i Seed, 2012. Page 76

Both, taro root and leaves are edible

on Honolulu's west side. By car, the trip takes no more than a quarter of an hour – at least outside the rush-hour, when Interstate 1 and Highway 92 can turn a motorway into something resembling a giant car park. Not only is Kalihi situated on the other side of Honolulu, it is also part of an entirely different world.

With a young KKV staff member as our guide we walk to a small community garden in the middle of one of the projects[3]. In Hawai'i's tropical climate, a lot of life happens outside and that makes poverty public and very visible. The housing stock is run down, and children play between rusty cars, overflowing trash cans and discarded mattresses. A few men sitting on folding chairs outside their homes eye us suspiciously. A barrier blocks off one of the roads. Next to it is a sentry post, and the set-up resembles a border crossing. 'That's a DEA[4] post,' our guide tells us. 'They regularly check anyone coming into the projects, and raids are common.' A little later we're back at the Roots Café. In Kalihi, it's mostly indigenous Hawai'ians who are living in the projects, says Sharon Kaiulani Odom, Roots Project Director at the Kokua Kalihi Valley Health Clinic (KKV). Migrants from Micronesia form another large group. They

---

3   US term for social housing
4   DEA stands for Drug Enforcement Administration

Sharon Kaiulani Odom at the KKV community health center

were forced to leave their homes after the United States military had used the islands to test nuclear bombs. Between them, the 200 KKV staff members speak 23 different languages. Lives in Kalihi are shaped by unemployment, poverty, drug addiction, and a high crime rate. And a lot of people suffer from chronic illnesses, says Kaiulani. She is a registered dietitian, and the 'Roots' project is not just an attempt to improve the health of the local community. Kaiulani wants to address the root causes by changing attitudes and helping people to be more self-confident and to take charge of their lives. She describes the vicious circle of poverty: if you are poor you cannot afford, a car and that often means unemployment and higher living costs. Public transport is almost non-existent in Honolulu, and most jobs are to be had in the tourism industry in Waikiki and remain out of reach for anyone in Kalihi without a car. Moving to where the jobs are is out of the question, rents on Hawai'i are extremely high anyway and in areas like Waikiki apartments are rented to tourists on a weekly or monthly basis, meaning affordable housing is just not available. And poor areas like Kalihi lack basic infrastructure – there are few shops with food mostly available in bodegas[5] and gas stations, where it is more expensive. Food on Hawai'i costs more than on the US mainland in any case, because 80 percent

---

5   In Britain called 'corner shops'.

of all goods come on container ships or have to be airfreighted. To get to a supermarket to buy in bulk and save some money or to visit a farmer's market for fresh produce is only possible if you have a car. Kaiulani Odom is absolutely determined to break this vicious cycle.

Roots Café at the KKV health center in western Honolulu

Tall and with long, free flowing blond hair streaked with grey, the native Hawai'ian cuts an imposing figure. Much of her skin is adorned with intricate, symmetrical tattoos, representing her ancestral lineage. Some have a spiritual meaning, while others are protective symbols or are connected to Hula, she explains. Kaiulani Odom dances Hula which in no way, shape, or form resembles the performance tourists enjoy watching over a Mai Tai and a BBQ. The traditional form of Hula narrates – through dance – Hawai'ian myths and historical events, and it is part of 'Aloha Aina', the love for the land. 'Food can connect us to our culture, neighbors, nature, air, water, and soil,' says Kaiulani Odom. In her opinion, the vicious circle can be broken by reconnecting with 'Aloha Aina', traditional culture, and the old Hawai'ian agricultural system, which could still feed everyone on Hawai'i. To her, being rooted spiritually and rediscovering what it means to identify as a Hawai'ian are part of the process.

Cooking classes are a first step. To change attitudes Kaiulani Odom and her team come up with creative solutions, and they have learnt to be patient. Take spam for example, still a food favorite on Hawai'i. Spam was invented by the US food giant Hormel. They added preservatives, sugar and salt to highly-processed pork and found that the mix can be canned and thus transported over long distances without need for cooling. And because spam is a highly-processed product, parts of a hog can be used that otherwise would be impossible to sell – nothing in the pinkish lump in the can reminds the consumer of the fact that pigs were involved in its making. During World War II, spam became a staple food for US soldiers. But while the enthusiasm for canned meat of dubious provenance waned quickly elsewhere, spam remains extremely popular on Hawai'i. What steak is to a Texan, a hot dog to a New Yorker or spaghetti to an Italian, spam is to a Hawai'ian. Spam, fried or grilled, spam with rice and fried onions, spam in soup, in sandwiches, with fried eggs for breakfast or as 'sushi', served on cold rice and wrapped in seaweed – spam is of such importance that it is celebrated in an annual spam festival. Spam contains a lot of fat, preservatives and sugar, and way too much salt, says Kaulani Odom. Many Hawai'ians are obese and prone to diabetes and high blood pressure which means spam should be off the menu. But at KKV nobody tells patients that they can't have spam anymore. Instead, they are offered a cooking class in which they can learn how to make spam themselves with fresh pork, organic sugar, and sea salt – cabbage or carrots are served as sides. Kids are welcome at the classes. They are the ones who will encourage their mothers at home to cook with fresh produce. Today, many Hawai'ians lack even the most basic cooking skills, says Kaiulani Odom. They will go to a shop to buy grated coconuts or coconut water because they don't know how to open a fresh coconut and hardly anyone uses coconuts in cooking. 'You are what you eat' is her mantra. Teaching it in antenatal classes, she tells the parents to be: when you eat good food, you lay the foundation for future generations. 'We change things one meal at a time,' she says.

Of course Kaiulani Odom knows that it costs more to feed a family with fresh fish, fruit, and vegetables than on a diet of spam and rice. KKV therefore works with 17 small organic producers – one of which is Dean Wilhelm's taro farm. Meat, dairy products, and fish are sourced from local traders. The produce is sold at the Roots Café, at a mobile market stand, and at several schools. 'Most people only occasionally buy organic produce from one of these outlets, and they may only choose a few items, but it is enough to start a conversation about food,' says Kaiulani Odom.

To Kaiulani Odom, Aloha Aina means to feed oneself and others by growing fruit and vegetables, cooking, and sharing meals. A few years ago, KKV rented 100 acres (about 40ha) of unused de facto wasteland from the council. Volunteers cleared the steep slope, terraced it, and transformed it into what is now the Kalihi Valley Nature Reserve Ho`oulu `Aina. The site is just a few miles from the health center. Visitors and course participants invite friends and neighbors along, and high-school students and youth groups bring siblings, parents and grandparents. Nobody owns the land, but all share the responsibility for it, the work, and the harvest. Ho`oulu `Aina is set up like a communal Ahupua'a system of old. Many Hawai'ian words have several connotations. Ho`oulu means 'growing' but also 'to facilitate growth', and 'aina' is a lot more than soil: 'Weaving together the cultural and social identity with the place where one resides, the Hawaiian worldview further understood 'āina not as simply land or property, but as that which feeds, signaling the inextricable relationship between land, people, and food. The values governing land and natural resources in Hawai'i ensured that the present generation would subsist totally from locally sourced food, and that the integrity of natural resources would be preserved so future generations could thrive. The fact that the residents of the Hawaiian Islands never relied on imported food well into the eighteenth century is a direct result of this integrated approach to the food system, land management and socio-cultural belonging'[6].

Things changed with the arrival of Captain Cook in 1778. During the decades that followed more Europeans and North Americans settled on Hawai'i, in particular sailors, missionaries, and traders. Initially the 'Haoles' were limited in what they could do because all land was communally owned and not for sale. But the white colonialists knew how to play different factions of Hawai'ians against each other, and in 1845, a change in the law allowed non-Hawai'ians to own land. The law reform, which is still known as the 'Great Mahele' ('to divide or portion') ended communal land ownership, and though it was meant to give secure titles to Hawai'ians, land ownership was also a prerequisite to plantation agriculture. A small group of traders and sugar barons still known as 'the big five[7]' managed to buy huge swathes of land or rent it cheaply. For the next 150 years, more and more sugar cane and pineapples were grown, processed, and exported, while at the same time, more and more food had to be imported.

---

6   Loc. cit. Pesticides in Paradise, page 10
7   Castle & Cooke, C. Brewer & Co., Alexander & Baldwin, Theo Davies & Co., und American Factors. (Source: Pesticides in Paradise)

Albie Miles is Assistant Professor of Sustainable Community Food Systems at the University of Hawai'i, West O'ahu. At some point, most of his students will do practical agricultural work. It merits a look into history to understand why it

On Maui, sugar cane was grown on large plantations. From the air, the devastation in the wake of a century of industrial agriculture becomes visible.

is particularly hard on Hawai'i to motivate anyone to become a farmer. After the 'Great Mahele', Hawai'ians had the right to just one percent of all land, says Miles. And to get access to it was very difficult: all claims had to be filed in English, a language most Hawai'ians didn't speak or understand. And applicants had to prove that they were living on the land they claimed, something that was mostly impossible for communal land. In general, Hawai'ians were only able to claim marginal land that wasn't very fertile, lacked water, or was part of one of the steep slopes, and therefore hard to access. This is the reason why even today, many indigenous Hawai'ians live on the dry, western side of the islands. Many other Hawai'ians had to move there during the land consolidation for the plantations. On Oahu and Maui, the relatively flat plateaus at the center of the islands proved to be ideally suited for establishing large industrially farmed monocultures. If not enough water was available, the plantation owners would build systems of interconnected canals and pipelines that channeled the water from the other side of the watershed across or through the mountains to the fields. The available labor force on Hawai'i was relatively small. The 'Big Five' therefore recruited workers – often whole families – from Korea, Japan, China, the Philippines, and Puerto Rico. Some came on time-limited work contracts and returned to their home countries. Others stayed, often not voluntarily but as bonded labor. 'By the beginning of the 1900s, the Commercial and Sugar Company had 25,000 acres of land, housed 3,200 workers, and returned a 20% profit to its stockholders (Kent 1983). At their peak in 1936, pineapple plantations occupied 89,000 acres across the Islands (Bartholomew et al. 2012), while sugarcane occupied over 200,000 acres as late as 1982. (...) As early as 1934-1936, Hawai'i was producing only 37% of its food, importing the rest'[8].

To maximize yields on the plantations, enormous amounts of chemical fertilizers were used, and the soils degraded quickly. Growing crops in huge monocultures increased pest and disease pressure, which plantation managers combatted by using vast quantities of herbicides and pesticides. In 1959, when Hawai'i became the 50th US state, wages started to rise. By the 1970s, companies like Dole and Del Monte started to shift their operations to low wage countries like the Philippines and Taiwan. In 2017, the last sugarcane plantation was closed. With the demise of the sugar and pineapple industry came a turning point for agriculture on Hawai'i: suddenly a huge amount of agricultural land became available and with water infrastructure in place. Crops could be irrigated if necessary. The question was who should be eligible to use this land in the future, and for what. Many Hawai'ians were hoping for a far-reaching land reform. If the land were to be parcelled off, small family farms would be able to produce all fruit and vegetables Hawai'ians needed. The politicians decided on a different course: they started a charm offensive, hoping to lure the big agrichemical companies to Hawai'i. The companies rented the land either directly or through the Department of Agriculture in Honolulu and its newly founded ADC development agency. 'In 2007, the ADC gave exclusive license to use, manage, operate, maintain, and control the infrastructure of the former Kaua'i sugar cane lands to a private entity called the Kekaha Agriculture Association (KAA). As part of this arrangement, KAA gained control of two critical ditch systems that represent much of west Kaua'i's agricultural water resources. KAA is primarily financed and run by its largest corporate leaseholders, which include Pioneer Hi-Bred (now owned by DuPont), Syngenta, and BASF; but it also receives sizeable annual management and project fees that are funded by taxpayers'[9]. The reasons

---
8   Loc.cit. Pesticides in Paradise, page 11
9   Loc. cit. Pesticides in Paradise page 14

Farming with Benefits

given by the state government were that the agrichemical companies would create jobs and contribute to economic growth. But in 2012, only 0.23 percent of workers on Hawai'i had a job with one of the agrichemical companies, and the number has dropped since. In the same year, the companies contributed a mere 0.52 percent to Hawai'i's GDP[10]. On the other hand, there are the costs for human health and the environment. And what hasn't changed is the fact that an island state with fertile soil and a climate that permits three harvests in a year has to import two thirds of all fruit and vegetables.

Hawai'i Congressman Chris Lee in his office in Honolulu

That has to change, says Chris Lee, Democratic member of the Hawai'i House of Representatives. By now he is in his third term. When he was first elected in November of 2008 he was only 27 years old and became the state's youngest congressman. As a true millennial, he communicates directly and using social media – we arrange date and time for the interview via Twitter.

We meet Chris Lee in his office on the 4th floor of the Statehouse in Honolulu. The wall behind is covered by a huge aerial shot of the district he represents in the southwest of Oahu. To him, climate change, agriculture and food security are the most important issues on Hawai'i. 'We have to survive as an island nation',

---

10 Figures from Loc. cit. Pesticides in Paradise, p 15f

says Lee. Global warming has increased the risk of hurricanes significantly. So far, Hawai'i has been spared, but if a hurricane like the one that devastated Puerto Rico in the fall of 2017 were to hit Hawai'i, it would cause significant disruptions. The food reserves on the islands are only sufficient for a few days, meaning that in an emergency, supermarket shelves would be empty within hours. But even more dangerous are coastal erosion and the rising sea levels. Within the last 20 years, 17 miles of beaches have vanished, and the erosion is speeding up, says Lee. 'For a long time, there was a conflict between environmental protection and private interests. Now hotels say: tax us more, but do something about coastal erosion,' he explains.

Warmer temperatures have led to an increase in invasive species, in particular mosquitos. And the warming of the ocean has fatal consequences for the coral reefs that protect the islands from the destructive force of the huge waves constantly rolling in from the Pacific. At the end of the century, the coral reefs will be all but gone. 'It's scary!' says Lee, in a way that is both direct and deeply personal. It's not a set phrase of a seasoned politician who discusses climate change and the consequences on an abstract level. Chris Lee has witnessed the changes over the past two decades, and like everyone else on Hawai'i, he lives with the consequences on a daily basis. When the beach you've gone to for your morning run becomes narrower and narrower until it disappears completely, climate change becomes a frightening reality.

Regarding Hawai'i's the survival as an island nation, Chris Lee believes the agrichemical companies are not exactly contributing to achieving this goal. 'Why do we produce seeds for export and not food? Why do we have the best land and three harvests a year but import 80 to 85 percent of our food?' To him, the central question is: 'How can this system be transformed into vibrant agriculture that feeds people, brings revenue and uses best practices that help mitigate climate change through carbon sequestration?.' He's worked hard to transform this ambition into political policy. In January of 2017 he introduced HB 1578. A 'Task Force for carbon and farming' is there to incentivise farmers and ranchers to improve the land's resilience, food security and the livelihood of people in Hawai'i while combatting climate change through soil health. The bill also includes a budget. 'The momentum is with us', says Lee. 'Everybody is behind the renewal of the food system, empowering farmers, providing economic benefits. Ten percent more local food would mean a huge increase in GDP and would help our transport goals because the food would not have to be flown in. Things are very different from five years ago. The anti-pesticide campaigns on Kauai and Maui really have helped. There is awareness now that didn't exist in

2013.' HB 1578 was passed, and only a few months later, on June 6th, 2017, the governor signed it into law, says Chris Lee. He's extremely modest about his success, but he's also proud of this achievement: to initiate such change, develop policies and find political solutions for the problems of the constituents who elected him – that's why he chose to go into politics in the first place. The Task Force is up and running, and apart from Chris Lee the team includes staff from the planning department and the department of agriculture, scientists, farmers, representatives from different farming organizations, and retailers. Only the agrichemical companies have no seat at the table.

Hector Valenzuela is a professor at the University of Hawai i in M noa. He works at the Department of Plant and Environmental Protection Sciences, and as an extension specialist, advising commercial farmers and vegetable growers, he trains extension agents across Hawai'i. 'When you start farming it needs to be set up for you, you need help and advice – that's what we provide,' he explains.

Plantation agriculture was pretty much finished when Valenzuela joined the university in the early 1990s. We meet him on a Saturday on the deserted campus in the northeast of Honolulu. The cafeteria is closed, and with no one in sight we are free to use one of the tables and benches outside. A thunderstorm seems to be brewing, but if it starts to rain we can retreat under the projected roof of the building. Valenzuela prefers to meet outside and face the weather rather than take us into his office. For good reason, his ideas about Hawai'ian agriculture and how small and diverse family farms could produce enough fruit and vegetables to feed everyone on the islands nearly cost him his job. 'Hawai'i and our university had world renowned research for sugar cane and pineapples, the horticulture department was seen as "ornamental", with one person for veg and one for fruit. Hawai'i was at the nexus between east and west, the Hawai'ian government wanted Hawai'i to be high tech and cutting edge and that included genetic engineering. The debate split the department,' says Valenzuela.

'My perspective has always been that we can deal with most problems using local, sustainable farming systems. That includes dealing with climate change. I like the science of GE crops, it teaches us a lot. But on the practical side: do we need to insert trait x to solve a problem? I cannot think of any specific trait that would achieve something that we can't achieve by way of other means. GE technology means depending on big corporations for solutions,' says Valenzuela. 'We can develop solutions for agricultural problems in the community.' This is an approach he pursues at the university and through the work in his department, much to the chagrin of the dean, who tried to revoke his tenure. But a commission found his research in line with his job description. 'On the one hand they left me alone after that, on the other I was ostracized like I was a snitch,' he recalls. In 2011/2012, he was transferred to a new department which for the last few years has been led by someone who, as Valenzuela puts it, is 'OK with sustainable agriculture and my research'. Nevertheless, he prefers to meet journalists outside, and not at the office.

And here's his vision for Hawai'i's agriculture and the future: 'All communities should be surrounded by green corridors where diverse, small-scale farmers produce the food for the community. The farms should employ production methods that increase soil health and maintain productivity.' Each farm is different, says Valenzuela. Farmers have to develop their own practices geared at improving nutrient cycles, increasing biodiversity, mixed cropping systems which complement each other, in their need for resources like water, light, and nutrients. Intercropping, mulching and composting should also be part of the system. For example, a four-acre farm could have one acre of intensive vegetable farming with rotation on individual beds, one acre of taro or sweet potatoes which are less intensive crops, one acre of forage for animals, and one acre with bio fuel. Hector Valenzuela and his team of extension agents conduct research, but also help with practical questions and advice. However, before the agricultural system on the islands can change, more Hawai'ians need to understand the connection between agriculture, food, and health, and more young Hawai'ians will actually have to become farmers.

So, let's get back to the narrow, forested Kalihi Valley northwest of Honolulu, and to the Ho`oulu `Aina nature and garden project run by Sharon Kaiulani Odom and the KKV health center. A narrow path leads to the traditional wooden structure which is open on two sides and serves as assembly point, meeting room, canteen, and workshop. On the day we visit, a seed workshop is in full swing; tomato seeds are being cleaned, dried and saved for the next season. Right now, everyone is on their lunch break, and Fiore Anderson has time to show us around. A few months ago, she got a staff job as a gardener at Ho`oulu `Aina, and for her that is a dream come true. She is in charge of the three-acre vegetable garden, a tiny part of the 99-acre nature reserve. Most of it is forest growing on the steep slopes of the valley. The Ho`oulu `Aina forest group works hard to reintroduce traditional Hawai'ian tree and plant species, such as the breadfruit tree, which used to be an important food source and could become one again. Eventually, all the imported tree varieties which were planted for lumber will be replaced by indigenous species, Anderson hopes. The trees on the level areas at the bottom of the valley were felled to make room for the vegetable plots which don't just need space but also enough light. There are many varieties of each crop, dozens of tomatoes, small, round egg plants, long, purple ones, and there is even a striped variety. In between grow different herbs and medicinal plants. When KKV took on the land, the soil was extremely poor, says Anderson, as the previous owner had mined top soil, filled sacks with it, and exported it to nurseries and garden centers on the US mainland. There used to be a well that fed two creeks but it dried up once the humus layer was removed. The gardeners and volunteers have worked hard to *improve* soil fertility. Once nature is in balance again, the water will come back, too, says Anderson. 'We try to make decisions that reflect what the plants want, not what we want,' she explains. And one day, they will plant taro. At present, there are only a few 'dry taro' plants; unlike the taro on Dean Wilhelm's farm, this variety does not need slow running water to grow in, but it doesn't taste as good, says Anderson. 'When the volunteers first come to Ho`oulu `Aina, most have never seen a taro plant,' she says. 'They don't know what fresh beans look like, how onions grow, and how many different herbs there are.' Thirty-two-year old Fiore Anderson sympathizes, as for a long time she didn't know either. She grew up in a poor, run-down part of Honolulu. 'At school I was a bit of a rebel,' she says. Rebellion was her way to deal with the problems at home and the health issues her parents were facing. Her mother is diabetic and also needs dialysis, while her father has a heart condition, and both have issues with alcohol. Sitting still in a classroom was hard for her: 'I hated school. I ended up in a special needs class, which was even worse, and eventually I dropped out.' Anderson was 18 when her daughter Sheylynn was born, and that made her think about the future and the fact that she wanted her child to have a different life. A friend suggested she should volunteer at the Aina garden. 'From day one it was a wonderful experience,' she says, being shown things and getting proper explanations, asking questions, being curious and learning – today that's normal, it is a way of life and something precious she would like to pass on, not just to her daughter, but to others, too, young people who are facing similar problems to the ones she once had. Fiore Anderson walks with us to a newly planted part of the garden. Tall trees provide semi shade, ideal for the traditional Hawai'ian medical plants which will grow here. Kaiulani Odom hopes that at KKV, Hawai'ian medicine will in future be practiced alongside western medicine. 'This is a Maile Hohone,' says Anderson, and points to a small, blue flowering plant. It can be used in cancer treatment but also for ulcers and gout. Next to it is a shrub with cone-shaped fruits that are rich in vitamins and good for diabetics. The Hawai'ian name for the breadfruit tree is Ulu; the fruits are full of proteins and minerals, and are very good for the immune system. It's the home of Haumea, the goddess of fertility,' says Anderson. 'The tree is female, as is the wind'. For Anderson, volunteering and working at the Aina garden was also an introduction to Polynesian-Hawai'ian culture. Today she feels a deep connection to those ancestors who first came to Hawai'i; she has found her roots and has become rooted.

The yields of fruit and vegetables at Ho`oulu `Aina are good, but the community project is not meant to be a market garden, the focus is on training and showing newcomers and volunteers what farming could look like. Most of the produce is used to prepare meals in the garden for anyone working there on the day, be it school kids, trainees, course participants, or volunteers. Anyone can take produce home, and the remaining fruits and vegetables are sold at the Roots Café. Ho`oulu `Aina is about raising awareness and about Aloha Aina – the connection between soil and food, health, and Hawai'ian identity.

Ma'o Farms trains a new generation of Hawai'ian farmers

If Hawai'i's food system really is to change, a lot of small family farms and market gardens will be needed to produce vegetables, fruit, grains, dairy, and meat. And that requires a new generation of young, well-trained farmers. The largest organic farm on Oahu is Ma'o Farms. Ma'o stands for Mala Ai Opio, and means both 'youth garden' and 'young garden'. The farm is situated in the Waianae District, on Oahu's west side. From Honolulu, a multi-lane highway leads past Pearl Harbor, a huge area with container terminals, cranes, and warehouses. The military facilities at the harbor are hidden behind fences and high walls. Once one has navigated the maze of exit and approach roads, traffic becomes light. To the right of Highway 1, the landscape is flat, brown, and desolate. On the left, every few miles an exit leads to a newly-built residential area or one of the shopping malls with well-known brands of shops and chain restaurants, each mall looking pretty much like any other. After an hour of driving, the state highway has morphed into a narrow road which veers north. To our left is a seemingly endless palm-fringed beach, on the right mountains rise up. Ma'o Farms lies hidden in a small valley. The soil here is dark, soft, and very fertile – if there is water. Without irrigation, it becomes as hard as concrete, says Hiwa Kanehaiua, who is responsible for management and administration. In 2001, her parents were among the founders of the 30-acre farm. Her father moved to Hawai'i from New Zealand, and her mother is a native Hawai'ian. Waianae is a very rural district. Many indigenous Hawai'ians live here, not necessarily out of choice. Waianae was one of the few places they could move to when they lost their land to the plantation owners. Originally, the whole valley was one big Ahupua'a system which provided food for a number of communities. But the plantation owner on the other side of the mountain range diverted the water shed to irrigate the fields on the central Oahu plateau. The plantations have closed down, but the military is still there and needs water, meaning it is unlikely that the canals and pipelines of this water infrastructure will be decommissioned any time soon.

Ma'o Farms works with a mobile irrigation system that can easily be moved from one field to the next. Good soil can hold more water, which is in itself a reason to use farming practices that help to increase the soil organic matter. On most fields, a number of different vegetables are grown, chosen to complement one another in their need not just for light, but also for minerals and nutrients and the microbes providing them. The multi-year rotation allows for a third of all fields to lie fallow. At least 30 different fruit and vegetable varieties are ready to harvest at any given time. Salads, scallions and fennel have a short growing period of just 30 days. They will be paired with varieties that have a longer growing period of 60 to 90 days. Some seedlings will be grown in pots and planted out later. An agroforest system shelters the vegetable plots from the constant wind coming from the Pacific, and protects the soil from drying out. Mango, lime, papaya, and citrus trees also provide an additional income, and are an ideal habitat for pollinators. Food production, soil improvement and education are the three goals Ma'o Farms is set up to achieve, says Kanehaiua. There are few schools and training facilities in Waianae District, and the job prospects are dire. Therefore, it is important to demonstrate to young people that farms and market gardens offer a lot of opportunities, 'You can grow good food for yourself and your family and you earn money'.

In cooperation with the local college Ma'o Farms offers a three-year trainee program. Only students from Waianae are eligible to apply. At present, there are 55 students, and for three days, they do practical work at the farm, while on the other two days, they attend college. Students do not have to pay for the course, unless they fail their exam – in that case they have to reimburse the full fee to the farm and the college, a rule that has proved to help student motivation enormously. Initially, the students earn 500 dollars a month, and in the second and third years they get a little more. And of course, everyone can take as much produce home as they need. At the end of the training, the graduates should not only be able to grow fruit and vegetables, but should also have gained the necessary management and marketing skills to run a business – from budgeting and pricing, to packing, distribution, presentation, and marketing to being computer literate and Internet savvy. Staff and the founders of Ma'o Farms get a salary, while any additional income through the sale of fruit and vegetables goes back into the farm.

Sorting and packing vegetables and salads is part of the Ma'o Farms training program

During our visit, only very few staff and students are out in the field. We've come on a delivery day and almost everyone is needed in the pack house. Entering the well-lit, freshly painted and immaculately clean facility it is hard to imagine that before Ma'o Farms took over the land, the building used to house thousands of broiler chickens. The atmosphere in the packing house is relaxed, and everyone seems to be in a good mood, singing along to the hits blasting from a sound system. In the morning, cases of freshly picked fruit and vegetables are carried from the fields to the back of the building, where the inside workflow begins: everything has to be sorted, washed, and weighed or counted and made into bundles. Different salad mixes have to be assembled, weighed, and filled into bags. On the wall at the other end of the building hangs a huge wipeable board, listing the orders and the produce available. Ma'o Farms delivers to restaurants, retailers such as Whole Foods, and it brings (bright yellow) veg boxes to private customers. The packers who are putting together the orders have to let those at the beginning of the chain know early which varieties have to be prepped first – each order has to be completed in one go and stacked in the cold storage room until the delivery van can be loaded. There is no room on the packing table for half-finished boxes that are missing a bag of salad or a bundle of carrots. In the adjacent office space hangs a wall chart that lists all orders in a manner that makes it possible to trace every delivery from the field to the customer. It is a complex system, and making it run smoothly depends on each team working fast, precisely and communicating well with colleagues. The time pressure is huge, as all boxes have to be delivered the same day. 'Students here acquire a bundle of skills which will serve them well not just in agriculture but in pretty much any job in the food sector,' says Lynn Batten, who's in charge of events, the farm to fork and school outreach program at Ma'o Farms. It often takes a lot of work to motivate students to join the course. For many parents, the education of their children is of no great importance, as what counts is a well-paying, steady job with benefits like healthcare. And few think about whether their food is healthy; for most Hawai'ian families, food needs to be cheap so that everyone has enough to eat. 'The students who come to our information days on the farm have often never tasted the vegetables that are grown here,' says Batten. 'The introduction day always includes food. We have to persuade them to try new things, and suddenly they realize that it actually tastes good! For many of them it is a real "food culture shock".' Lynn Batten serves as a mentor throughout the course. She not only knows every student's name but also learns a lot about their family backgrounds and the circumstances under which they live. 'Throughout the course, we need to keep asking ourselves whether we have created conditions that make it possible for each student to actually learn,' she says.

For our final conversation we sit on the roofed patio. It's a beautiful space, used for meetings and the 'farm to table' evening, when a celebrity chef takes Ma'o Farms produce to create a meal for well-heeled guests. Events like these are needed to thank existing sponsors and, hopefully, gain new ones. It's midday and really hot outside. On the patio, a cool breeze carries the scent of the herbs and edible flowers growing in pots nearby, and from here, one has a magnificent view of the whole farm and the mountains beyond. Next to the covered area, raised beds will be created and planted with herbs and vegetables. In future, visiting schoolchildren will be able to see them close-up and get a taste. Some of the ideas came from the former First Lady, Michelle Obama, who visited the farm twice. The first time was in 2012, after the release of her book[11] about the vegetable garden she established in the grounds of the White House, and the second time, the First Lady came 'just because she likes it here', says Kanehailua. It's easy to understand Michelle Obama's enthusiasm for Ma'o Farms and its educational concept: every conceivable aspect has been thought through, and everyone tries their utmost to help the students. But it's still a long way for the Ma'o Farms' vision for a sustainable food system on Hawai'i to become reality. At present, the produce does not go to Hawai'ian customers in the neighborhood but to organic supermarkets and high-end restaurants in Waikiki. These cater for tourists who expect farm-fresh organic produce because that's what they are used to at home.

Hawai'ian grown fruit and vegetables will only become available everywhere once becoming a farmer or a grower is a realistic option and a good career choice for young Hawai'ians. 'At present, young farmers still face a lot of hurdles,' says Albie Miles, professor at the University of Hawai'i. 'And farmer training programs are only now being developed.' The biggest problem is the lack of available agricultural land. The plantations may have been closed down, but land, water rights and infrastructure are still owned by corporations and rich individuals. In view of the continuing boom in tourism, it makes sense to leave fields fallow and hope for land-use changes which might allow parcels of land to be sold off to developers. Agricultural land usually is only available on short-term leases. Under such circumstances, planning becomes difficult, and investing makes no sense. Farmers cannot build on agricultural land, which means they cannot live on their farm, and apart from long, costly commutes, theft often becomes a big problem: a mango orchard full of fruit might be 'harvested' by thieves overnight. On Hawai'i, pretty much everything is more expensive than on the US mainland,

---

11 Michelle Obama: American Grown: The Story of the White House Kitchen Garden and Gardens Across America, Crown, 2012

says Albie Miles. Energy costs are the highest of any US state. High transport costs add to the price of chemical fertilizer, machinery, and building materials. Because agricultural land is scarce, rents are also high, and fruit and vegetable growers on Hawai'i have to compete with farmers in California who farm huge acreages and can produce cheaply. And it's hard for young farmers to get low interest loans, which means that starting even a small enterprise requires taking on a lot of debt. Some basic parameters need to change for a sustainable, local food system to emerge on Hawai'i, says Miles. The availability of cheap loans and the right to live on the land one farms is one. But a processing infrastructure and a variety of farmer training programs that are rooted in practical farming rather than academic achievements are also necessary[12].

Every year, more than seven million tourists come to Hawai'i, which itself has a population of just 1.3 million. Money from the tourism industry could pay for subsidies. One idea would be to create a quota of food that has to be purchased from local, native Hawaiian, and organic farms, says Albie Miles. It would force hotels and restaurants to source at least some fruit, vegetables, dairy and meat products locally. And there should be incentives for farmers who use regenerative agricultural practices and improve the soil. In his opinion, 'carbon farming' should be an additional revenue stream. Hawai'i may still be a long way away from a sustainable food system, but the first steps have been taken. The 'Hawai'ian Renaissance', with more Hawai'ians reconnecting with traditional practices and knowledge, has helped many people to understand the link between sustainable agriculture and healthy food. Aloha Aina stands for the knowledge that the islands can provide enough food for all. Whether it's in the 'Roots' Garden near the KKV health center or on Ma'o Farms, the enthusiasm for sustainable agriculture and good food amongst young Hawai'ians is palpable. Not only on Maui did Hawai'ians like Alika Atay get involved in the anti-pesticide campaign, with the fight against the big agrichemical companies becoming the motivation to stand for public office. Atay was elected council member in 2016, but his main job is still being a farmer. And he advises young Hawai'ians on Maui, Oahu and the Big Island how to overcome obstacles and start their own farm. On Maui, a political fight over the future use of plantation land has begun. Alika Atay and his colleagues insist that at least part of the land will be made available to young farmers who want to grow fruit and vegetables for Hawai'ians. These days, such initiatives can hope for support from the government in Honolulu. Chris Lee's climate change and sustainable agriculture Task Force is just one example.

---

12 Ma'o Farms is a unique organisation that provides excellent training, but only for students living in the distric.

In our conversations, one farm was mentioned time and again: Moloa'a Organica. It's the farm of Ned Whitlock, situated in the north-easternmost part of Kauai on what used to be a plantation. Initially, sugar cane was grown here, then pineapples, and finally papayas. In 1981, a storm caused such devastation that the land was fallowed until 2000, when individual land parcels were sold off. Today, it's the biggest piece of continuous agricultural land on Kauai, and is mostly farmed organically. The attempt to find a farm in the land area extending to the east of the main road all the way down to the ocean requires a detailed description and a good eye for landmarks. The deep ruts in the dirt track slow driving down to walking speed, and the washer system turns the red dust on the windshield into a sticky paste. But it is not just the impaired visibility that makes orientation difficult. Each crossing looks like one we have passed before. In fact, everything looks the same: the shrubs and trees that colonized the land after the storm are all at the same height and have grown into a dense, green wall that completely hides the farms from view. Finally, we find the right turn-off, and after a few hundred yards, we stop in front of an open packing shed, next to a tractor that has seen a lot of field work. To our right, we see long rows of vegetable beds and huge, fruit-laden trees. We've found Moloa'a Organica. Ned Whitlock, tall, lean, and with deeply tanned skin, comes to meet us. In shorts, faded shirt, work boots, and wearing a floppy hat, he has something of a beat-generation hippie about him. And maybe that's what he is. Whitlock was born and raised in Arkansas. He studied biology, went to New Zealand for a year on an exchange program, and decided to make his way back through Asia, staying in India for year. Back in Arkansas, he opted for farming the land his father, an oil executive, had bought in the Ozarks. The climate in the Ozarks can be rough, and that's not just the weather – organic farmers seem to be something of an endangered species there. 'My wife and I decided that we wanted to live somewhere warm and near the ocean,' is Whitlock's only comment on farming in the Ozarks. In 2000, they sold the farm and shipped the tractor and all that would fit into containers to Hawai'i. It took two years until they found suitable land to buy. They invested the last of their money in an irrigation system; summers on Kauai are extremely hot and dry. The farm has 28 acres (about 11 ha), but only four acres (1.6 ha) are used for vegetable growing, with fruit and lumber trees growing on the rest of the land. And rows of trees grow between the long, narrow vegetable beds. 'Trees are at the center of everything we do on the farm,' says Whitlock. Trees are ideal for pollinators and other beneficial insects, they provide shade, enhance soil quality, and the fruits are a good, year-round income source. Throughout the year, 50 to 60 vegetable varieties are grown on the farm. Ned Whitlock has spent a lot of time devising rotations and cropping plans. Now, dragon fruit climbs on nitrate-providing legume trees which also provide shade for the beans in the adjacent vegetable bed. At regular intervals, Whitlock has planted neem trees; passing a tree, humans don't notice much of an odor, but pests give the leaves and the surrounding area a wide berth. Because the tropical climate allows for three planting seasons and harvests, what would normally be a three-year rotation is condensed into one. The vegetable beds on Moloa'a Organica seem like lush clearings in a tropical forest. 'But soil fertility remains a problem,' says Whitlock. 'even after lying fallow for 20 years with 15 years of organic soil management, we still haven't made up for 150 years of plantation monoculture.' There is no method for soil improvement that hasn't been tried out on the farm. Comfrey tea is brewing in large drums. A wormery provides worm compost, and Whitlock makes his own biochar. He has also modified the method of compost production developed by the Korean Cho Han Kyu, who inoculates compost to produce a liquid fertilizer. For the inoculant, he makes a starchy paste from potatoes, cassava, or taro, then he adds virgin forest soil and sea water or salt and water. The mixture starts to ferment after 24 to 36 hours, and it will then be filtered, diluted, and fed into the drip irrigation system. Twelve staff work on the farm part or full-time. After the harvest, fruit and vegetables have to be washed, sorted, weighed, and packed. About half of the produce goes directly to wholesalers and major customers like hotels, while the other half is sold at farmers' markets in the area. Among the customers, there aren't just tourists and Haoles, but also many indigenous Hawai'ians.

On the farm, subtle and not so subtle signs of climate change can be found everywhere. The most obvious change is happening with the mango trees, says Whitlock. Normally they flower in spring and the fruit is harvested in summer. We visit the farm in October and many of the trees are in full bloom. And it's become a lot warmer, says Whitlock. Even in winter the water in the ocean isn't cold anymore.

I ask him what his vision is for the farm. Very soon, all energy needed on the farm will come from solar power, he says. And a bit more land would be great. It would allow him to work with longer rotations and grow additional biomass for mulching and composting. Right now, 20 acres (8 ha) are available for sale nearby – for 1.5 million dollars. No farmer can afford land at such prices. Probably someone will buy it to build yet another luxury villa, says Whitlock. In his opinion, Hawai'i has enormous potential, and it would be easy to produce enough food locally to feed not just Hawai'ians but the seven million tourists that come each year, too. And on his farm, there is also still room for improvement, he says – growth periods, carbon and water cycles, soil organisms…. 'We need to get all these systems talking to each other'.

In Honolulu, Rebecca Ryals investigates how systems communicate with each other. She has studied soil science at the University of California, Berkeley and did her PhD with the Marin County carbon project, which demonstrates methods of carbon sequestration on ranchland. What are the options to sequester carbon through agricultural practices? How do they work? What additional benefits does a higher carbon content in the soil provide for plant growth, water storage, and recycling of nutrients? And are there potential trade-offs and unintended consequences? These are some of the questions she tries to answer for Hawai'i. In California, she was used to working on standardized test plots, but because of its unique climatic conditions, soil on Hawai'i reacts very differently to similar soils on the US mainland. So far, little basic research has been done, and tools like accurate soil maps that are available for most of the US don't exist for Hawai'i. An app designed by Colorado State University allows a farmer to enter GPS coordinates of a particular field which connects to a soil map for the region. He can then enter additional information about crop rotation, crops planted, inputs, etc., and get a projection of what the likely outcome would be for certain practices. None of that is transferable to Hawai'i, with its unique climatic conditions and volcanic soils.

Ryals therefore is trying to figure out the basics: what do we know about soils on Hawai'i? Along with her students, she conducts research on Maui and the Big Island, which has a lot of grassland and ranches. 'Ranchers have a deep appreciation of the soil but little knowledge of what good soil actually is and what it means,' says Ryals. The first phase of the research project is about gathering as much data as possible. The team chose a large farm near Maui's National Park. The land is very varied, ranging from very dry to extremely wet. In the summer of 2017, Ryals and her students took more than 1,000 soil samples at depths of 15cm, 30cm, 50cm, 75cm, and 100cm. They also recorded the exact location, elevation above sea level and visual observations like tilth, structure, soil type (silt, sand, clay…), and surrounding vegetation. We walk over to the lab where the soil samples are stored in boxes, ready to be analyzed for their carbon content. The most expensive part of the procedure is the preparation: each sample has to be air or freeze dried. With tweezers, remnants of roots have to be picked out. Once the samples are 'clean', they are pulverized, wrapped in aluminum foil, and run through an 'elemental analyzer' which measures the CO_2 content. Once all samples have been analyzed, the team will know how many samples are needed to get a reliable picture, providing enough detail to evaluate changes. In phase two, the team will collect soil samples on the Big Island which hopefully will provide information about soil quality and the potential for carbon sequestration – which might be huge, given the volcanic origin of the islands.

Rebecca Ryals is optimistic, not just for Hawai'i, but for agriculture globally: 'Carbon sequestration is one powerful tool to reduce the temperature of the atmosphere. It is immediately available if there is the political will. Scientific research, study after study confirms: Yes, it can be scaled up. Yes, it will have a big impact on the climate and if we don't do it, it will cost us dear because we will need more and more input to get the yields we need.'

Profits on the Prairie

'Japanese Quarter Horse'

Mark Frasier kneels down and carefully spreads a bunch of blue grama grass to expose the center. The narrow blades of grass stand about 8 inches (20 cm) tall. Now, at the end of October, it looks dry, brittle, and brownish-yellow in color. It's tightly bunched and that is a good sign, explains Frasier: 'If there is a bald spot in the center, we haven't left the cows out here long enough; when the grass is green, it needs to be grazed down far enough for the young shoots in the center to get enough light and grow.' Individual tufts of blue grama grasses stand tall in an otherwise dense blanket of very short, wider blades of grass. Even some green, new shoots are visible. 'That's buffalo grass,' says Frasier. 'It spreads through a rhizome, and therefore grows evenly.' Wherever we stop, he kneels or squats because only at ground-level and close up is it possible for the diversity of grasses and plants to reveal itself in what otherwise looks like a vast, monotonous expanse of beige, light brown and yellow, typical for short grass prairie in fall.

The 44,000 acre (1,700 ha) Frasier Ranch is situated in eastern Colorado, just a few miles beyond Last Chance (population 23), at the crossing of US Highway 36 and State Highway 71. Two brick posts with a wooden roof mark the entrance, with 'Frasier Ranch' spelt out in wrought iron letters. A dirt road leads onto the ranch and vanishes in this far-reaching landscape and the vastness of the horizon. The farm buildings are sheltered in a dip and hidden from view. But we pass the farm and drive towards a ridge visible in the distance. When we get out of the car, it is quiet except for the steady wind and the rustle of the brittle vegetation, adapted to frost, heat, and drought. The prairie grasses seem to cling to the ground; only the seed stems stand about a foot tall, blowing in the wind like tiny flags. Above us is the expanse of a vast, deep blue sky – it could be the set for a Western; I'm half expecting a rousing score with riders appearing on the horizon.

Mark Frasier inspecting grass growth on his ranch

On the ranch much is very similar to a time some 200 years ago, before the first settlers arrived, and when millions of buffaloes still roamed the prairie. On the Frasier Ranch, they have been replaced by cattle: Angus, Hereford, and Simmental crossbreeds. The modern cowboys have swapped horses for quad bikes, occasionally referred to as 'Japanese quarter horses', depending on the brand.

Over millennia, grasslands and grazing animals have evolved together – in African savannahs and the steppes of Asia, as well as on the pampas of South America and the North American prairies. The less rain that falls in an area, the shorter the grasses will be that grow there. Until recently, the 100th degree longitude separated the Great Plains between the Mississippi and the Rocky Mountains into a wetter, eastern part, and a much drier western side. Because of the changing climate, this watershed has now shifted to the 98th degree longitude. To the east, severe rain events have increased markedly, while the west is now in a continuous state of drought. The National Oceanic and Atmospheric Administration (NOOA) publishes a monthly drought monitor. In the western United States, large areas are marked yellow, orange, red, or dark red, with yellow indicating abnormally dry conditions and dark red exceptional drought.

Farmers can't influence the amount of rainfall on their farm, but they can definitely change the water infiltration rate and the water storage potential of their soils. As we saw in chapter 1, both are direct functions of soil quality. In generally dry areas like the short grass prairies of the western US, the plants that naturally grow there need less water and must be more drought-tolerant. But such ecosystems are finely tuned to maintain their balance, and even small disturbances can have far-reaching effects. The Zimbabwean ecologist and rancher Allan Savory was one of the first to describe the immense importance of grazing animals like buffalo and cattle for the maintenance of grasslands like savannahs and prairies. His research led him to believe that both over- and under-grazing can damage grasslands. He went on to develop a method for the 'holistic management' of cattle which can 'revive' even severely degraded soils. The journalist and author Judith D. Schwartz visited Savory at his research station in Zimbabwe and wrote[1]: 'One useful way to understand Holistic Management is that it addresses a basic challenge in seasonally dry environments like Zimbabwe: how to maintain moisture in the soil from the end of the rainy season to the beginning of the next. Without animals eating the grass and pressing drying vegetation into the ground, plant matter accumulates without breaking down. This blocks fresh growth and inhibits biological decay. The result: the plant material oxidizes and the soil loses carbon and water as well as the capacity to support plant and animal life. In other words, the synergy between the land and the vegetation and the ruminant's digestive process can forestall, and even reverse, desertification in grassland ecosystems.'

Mark Frasier can still remember the day more than 30 years ago, when he first heard about holistic management in a lecture given by a member of the US

---

1 Judith D. Schartz: Water in Plain Sight. St. Martin's Press, 2016 p.18

Savory Institute. 'It was a total light bulb moment,' he recalls. 'I was convinced and I saw on the ranch that what I heard was correct.' Some of the ranch land had been grazed far too hard. The spring rains turned the soil into mud which was literally baked into bricks during the summer heat and deep fissures and cracks developed. Other parts of the ranch were under-grazed. There the grama grass tufts had developed the typical bald patches in the middle with no new shoots coming up because they didn't get enough light and were 'smothered' by the dying biomass covering them.

After attending the Savory Institute lecture, it wasn't just the economic implications that convinced Frasier things on the ranch needed to change, his decision was also influenced by family history. His great-grandfather homesteaded in Nebraska. His grandfather was an excellent businessman who got his family through the Great Depression and the Dust Bowl by trading cattle, says Frasier. In the 1940s, he bought a ranch in eastern Colorado. Soon, he started to see his neighbors once again plowing up the grassland to plant wheat. For his grandfather, the memories of the dust storms, the poverty and despair, were still raw, says Frasier, and now he saw it happening all over again. In 1946, he bought land in eastern Colorado which had never been plowed and established the Frasier Ranch, which developed into the business Mark Frasier and his brother are running today.

Frasier studied agricultural economics and when he came back to the ranch he started to focus on the business perspective. After listening to the Savory Institute lecture, it made sense to him that cattle on degraded grassland will put on less weight and will take longer to do so. Holistic management seemed like the perfect solution, and Frasier wanted to introduce changes fast. His father had not only erected a number of barns, but also had several farm tracks built, so that every section of the ranch became accessible. Half of the ranch has no water, while on the other half there are a number of wells, but not all are good. Frasier's father had 45 miles of pipelines laid so that the cattle could get water everywhere from the same good source. So, pretty much everything was in place to switch to holistic grazing; all Mark Frasier needed to do was set up fences to create paddocks, an investment his father considered an utter waste of time and money. With hindsight, Frasier thinks he introduced too many changes too fast. 'I didn't take the team with me,' he says. In particular his father couldn't see what he was trying to achieve. But once he agreed to attend a holistic management workshop and training, he, too, was convinced by Savory's approach.

Blue grama grass

We are visiting on a sunny but crisp fall day and under the deep blue, cloudless sky; the expanse of the prairie is of such stark beauty that I want to look up and out towards the horizon. Mark Frasier, however, is completely focused on the ground and reads it like a tracker: where are the hoof imprints, where have the animals grazed too much, and where did they not go? Holistic grazing isn't a system that can be put in place and then remains static. It's only through constant attention and adjustment that the balance between animals, plants and soil can be maintained, a balance that continuously shifts depending on the weather and the season. To get it right, one needs a lot of knowledge, experience, and the patience to observe. In this semi-arid region, plants have to make use of any available moisture, be it morning dew or a rare rainfall event. Frasier sees grasses as indicator plants: on sandier soil, you find lots of sage brush, he explains, as it helps to protect the soil from erosion through wind. The ranch mostly has loamier soil, and that is good for short grasses. Most plants have a narrow window for growth, and they depend on precipitation during that time. Prairie grasses are more flexible, and they bide their time, but once it rains they have to be off the mark very quickly. Blue grama grasses can develop seed heads between May and September, and they will do so in record time as soon as there is rain. Buffalo grasses, too, are survival artists. Shortly before our visit, temperatures dropped to 20F (-7 °C) for several days, but Frasier finds fresh, green shoots.

Since grassland and grazing cattle are in synch once more, the landscape has changed. After a rainstorm, Frasier now sees deep ridges in which clear water flows rather than the wide shallow beds with slow running, muddy water – run-off from the neighboring farm. 'We caught quite a bit of their land,' says Frasier. Now there is more grass, it catches the soil and narrows the channel in which the water flows until it finally drains away.

Does he see signs of climate change? In 50 years on the ranch, he hasn't seen any permanent changes but a lot of extended weather periods, some wet, but mostly too dry, says Frasier. For ten years, from 2002 to 2012, the annual rainfall remained significantly below average and the whole area was in a state of continuous drought. 'It was very disheartening out on the ranch,' Frasier recalls. 'We got by, but it felt like you made no progress, you just tried not to go under. We just continued doing what we were doing because we thought we were on the right path. In lean years, you can actually see the result of your management. Holistic grazing got us through the drought.'

The grassland is intensively grazed for a short period, followed by an extended recovery period. 'The plant community needs to be as diverse as possible,' says Frasier. 'Once the rain returns, the ranch will recover more quickly. We are always ready for the next rain so that you can catch as much of it as you are able to.'

The ranchland is divided into 125 pastures, varying in size from 250 to 300 acres (about 100 to 120 ha), depending on their location and grass quality. Frasier shows us a grass patch close to the farm track. There are a few bald patches where the soil is exposed, leaving the surface hard and cracked. If cattle were to graze intensively here, their hooves would break up the crust, and during the next rain, water infiltration would improve. While grazing, the cows would of course deposit numerous cowpats – perfect fertilizer that would boost plant regrowth. The result of intensive grazing would be a denser grass cover. That, at least, is the theory – which unfortunately does not translate into reality on the ground.

To get an even result, large numbers of animals were kept on this relatively small pasture for just two hours. But instead of grazing thoroughly and leaving cowpats, the animals just stood around waiting until they were allowed onto the next pasture. After he introduced holistic management on the ranch, Frasier spent the next ten years putting up fences and making mistakes. And the following decade was all about trying to balance the system. 'Each season is different,' he says. 'Taking the right decisions under continuously changing conditions has been a long learning process. Now I watch the livestock and let them tell me. If they don't react the way they normally do they are giving me a signal, they tell me something is wrong.'

Everyone hates branding cattle but in Colorado it's the law

Even small topographical differences can be of enormous importance. Frasier points out the bottom of a slight dip where moisture can collect. Now, at the end of October, it's just enough for sugar-rich cold season grasses to grow. During the summer months it's far too hot, they just dry up. Regarding nutrients, they come with a kind of 'best graze before' date, and that needs to be taken into consideration when the paddocks are being allocated. How long the animals stay on each pasture has to be carefully planned in advance, too. Cold season grasses shouldn't grow too tall, while warm season grasses should never be grazed too hard. How much feed is available in any given year depends on the weather. If there is a long phase of gentle spring rain, it will likely be a good year. But after a long drought, not even prairie soil can absorb a summer downpour with two inches of rain in 45 minutes. Which is not a problem as the water will collect in the dips and slowly percolate into the aquifer below. But to the grasses, this moisture is lost as even their deep roots cannot reach it.

Although things may be changeable and unpredictable, a ranch is a business and has to be profitable. Mark Frasier achieves this goal by way of marketing: part of the meat he produces will be sold under the 'all natural' label. If the ranchers meet the strict conditions and standards, the meat will be sold to consumers under the label and for a higher price. The other animals Frasier raises are sold into the conventional market. Working with two production systems gives him enough flexibility to react to weather and market conditions.

The 'all natural' label guarantees that the cattle have spent 75 percent of their lives on grass, that they were never treated with antibiotics, and that the additional feed they receive in winter does not contain any animal fats. Frasier works with a number of small cow-calf producers. The calves are born in spring and stay in a herd with their mothers for seven months before they are weaned. Only then do they come to the Frasier Ranch where they will stay through the winter and the following spring. At some point between July and October, depending on their weight and the trends in the meat sector, Frasier will send them to a finishing unit where they will stay for about five months. When they reach their slaughter weight, the animals are between 24 and a maximum of 30 months old. The finishing unit is a small family enterprise, just 20 miles (about 35km) from the ranch. A maximum of 7,000 animals are fed wheat grown on the farm. For comparison: the capacity of a large feedlot in the US is 100,000 cattle and more. Mark Frasier can sell almost all of the meat he produces under the 'all natural' label to the Whole Foods supermarket chain, recently bought by Amazon.

The winter months on the ranch are critical. Adult animals can put on weight from shortgrass prairie alone. They can eat vast amounts of roughage which calves can't digest as their stomachs are not yet developed enough. They need supplementary rations consisting of sugar beet pulp, alfalfa, hay, and corn.

During the summer months, the ranch can not only sustain far more animals, but soil quality and biodiversity profit from well-managed, intensive grazing. The holistic management of cattle on prairies can have the same beneficial effect the roaming herds of buffalos used to have.

In spring, Mark Frasier therefore buys in yearlings which he keeps on grass over the summer before he sells them on to a feedlot in fall. How many animals he buys in depends on the weather: in a year with sufficient rainfall, he will buy more yearlings than in a dry one.

To work with two different production systems is complicated and needs a lot of planning, but the resulting flexibility is often key to the profitability of the ranch. Twice a year, in spring, when he buys yearlings for conventional meat production, and in fall, when he brings in calves for the 'all natural' program, Frasier has the chance to react to weather and market conditions. During harsh winters or very dry summers, ranchers who cannot vary stock density have to buy in feed just when it is scarce and expensive, or they need to sell off animals even if the price is down.

Even though Frasier has studied agricultural economy he looks well beyond the profit the ranch can make. He is convinced that the economy in rural areas should function like an ecosystem – a close network at local level. The 'all natural' label and the higher price he gets for the meat allow Frasier to build reliable, long-term relationships with local businesses: from the cow-calf operations and the feed suppliers to the finishing operation and the processors. What matters is quality and continuity – and both can only be maintained if everyone in this system has a sustainable business. 'It's a system that needs a certain number of workers and creates new jobs, and that is very important,' says Frasier. There are already a lot of small communities in eastern Colorado that have no school, no shops, no restaurant, and no bank. Every new job in such communities will create more work opportunities: workers have families, they need to eat, they need places to live and shop, the kids need to go to school, cars and appliances have to be serviced and repaired – and suddenly there is enough demand to open a school, run a business, a restaurant, a repair workshop, or a bank…

But regarding the conventionally produced meat, Frasier, too, has to see where he will get the best price, and that usually isn't nearby. He buys most of his yearlings in the neighboring states of South Dakota or Nebraska, and at the end of the summer, he will sell them to one of the big feedlots. Conventionally produced animals reach their slaughter weight in just 18 months – any profit doesn't come from the quality of the meat, but from a large number of animals being ready for slaughter in as little time as possible.

On the day of our visit, 350 calves have just been delivered. They will spend almost a whole year on the ranch. The welcome committee consist of several mother cows who lead the youngsters towards the feeder wagon – after the journey, munching on some hay will make settling in a lot easier. Next come the registration, a health check, vaccination, and branding. It's a job everybody on the ranch hates, but in Colorado, branding is still required by law. Farm manager Jake, Mark Frasier, and his brother Chris are used to working fast and as a team: four or five calves wait behind a partition, the first animal in line moves into the chute and is fixated. With a quick feel, the farm manager establishes whether he is dealing with a female calf of a young bull, then he places the red-hot branding iron on the animal's flank and vaccinates it, while Mark Frasier fixes the ear tag and enters the registration number and gender into a tablet computer. The whole procedure takes no more than 40 seconds, then the gate opens and the calf races to the end of the coral where the other newly branded youngsters are already munching on some hay, seemingly having forgotten the ordeal. It will be evening before all 350 calves have officially become part of the Frasier Ranch herd.

Before we leave, I ask Mark Frasier about his plans for the future. It has taken him 20 years to re-establish a balance between soil, plants, and animals on the ranch. Now it's all about maintaining the shortgrass prairie as a functioning and profitable ecosystem, not just for himself and his family, but for future generations.

A few days later, we visit an organic farm in Nebraska, 300 miles (about 500km) to the east. Kevin Fulton is waiting for us outside the farmhouse. It is another beautiful, sunny fall day, above us the expanse of a vast blue sky with a tell-tale stripe pattern: we are definitely in fly-over country. At least half a dozen planes are visible, making their way east or west, each leaving a thin white condensation trail that slowly becomes wider until it finally disappears. It's rare to experience the discrepancy between the noise, the traffic, and congestion of a big east or west coast metro area and rural America as clearly as we did that morning. Nebraska has a population of just under two million – about the same as the city of Hamburg, just spread out over an area half the size of Germany.

While the planes keep drawing their patterns in the sky, we are trying to make our way to the highest point on the farm. Kevin with his daughter Cammy on the quadbike, her older sister Coleen, her younger brother Tim, Martin, and I on foot. Half of the 800 acre (325ha) farm is native mixed prairie. This part of Nebraska is a transition area between shortgrass prairie in the semi-arid west and longgrass prairie further east, where the annual rainfall is much higher. The prairie we are walking through is very different from what we saw on the Frazier Ranch: there are fat bunches of buffalo grass, and blue stem, Indian and switch grasses reach up to our thighs. Kevin Fulton is trying to locate the 'dug-out', the remains of the shelter in which the settlers who first claimed this land lived. Learning about local history is one of Fulton's passions, and it takes him just a few minutes to find the spot where the Thompson family once lived: a slight hollow, dug into the slope of the hill and long since overgrown by prairie grasses. The first settlers who came here during the second half of the 19th century had no building materials; there were neither stones nor trees that could be felled. Anyone living too far from a railhead or too poor to afford the supplies that came on the goods trains had no option but to dig a hole in the ground and use grass sods to build the walls and a roof. A bit further down the hill Kevin Fulton shows us the well the Thompsons once dug. He doesn't know what happened to the family, or when they might have moved on to find a better place to run a homestead.

The Fulton family standing in the remains of a dugout Kevin discovered on the farm

From the other side of the hill, we spot some of Fulton's cattle, brown, black and white dots, slowly moving along. He has fenced off 30 pastures, the 150 Galloway cow calf pairs graze each for one or two days, before he moves them on to the next. The whole herd is between 600 and 800 head strong. Originally, Galloways come from Scotland; they have a dense, sometimes slightly curly coat which gets them through the winter. Some are white, which is really great in the summer heat, says Fulton, and they are known to have excellent meat quality. Galloways are shorter than Herefords or Angus, resilient and undemanding – while other breeds are choosy and graze some grasses and herbs while avoiding others, Galloways munch through everything the pasture has to offer. One aspect of holistic grazing or high stock density grazing is to only let the animals onto a fresh paddock once they have evenly grazed the old one. A choosy breed will leave the plants they don't like, meaning those will come to seed and propagate. Over time, the plant diversity decreases, and when the animals come back the following year, they will find more of what they don't like.

'Modern beef cattle have forgotten how to graze,' says Fulton. 'Modern Angus expect a feed truck to show up every day.' On the dry and brittle shortgrass prairies, high-density stock could easily cause a lot of damage in a relatively short period, but mixed prairies with their much taller, denser grasses actually benefit. Fulton has put up 15 miles (almost 25km) of permanent fences and uses movable fences to create paddocks that give the cattle access to one of the 15 water sources on the farm. He also keeps a flock of 500 sheep. Their way of grazing complements the grazing patterns of cattle.

With holistic grazing management on 400 acres (160ha) of prairie, Fulton can keep three times as many animals as his conventionally farming neighbors. The Galloways stay on the farm until they have reached their slaughter weight. Certified organic meat from grassfed animals should sell for a really good price – except that Fulton's farm is situated in central Nebraska and no certified organic slaughter facility is within reach. For a long time, selling meat from grassfed animals was good business, but then import regulations changed. Suddenly, big meat processors could buy cheap grassfed beef from Argentina and other South American countries, while prices for grassfed meat produced in the US collapsed.

Grazing native prairie - cattle on the Fulton Farm

Fulton is a rancher at heart who would love for all his land to be nothing but native prairie and grassland, but in order to stay profitable he has to grow corn, too. In the last few years, selling beef with a profit has become so difficult that he has even plowed under a few acres of prairie. 'My heart really bled, but I can make $1,000 net profit on organic corn,' he says. 'With beef that's impossible as profits on cattle have halved over the last few years.' Because of the depressed prices Fulton has reduced the number of cattle and has had to make hay because there weren't enough animals to graze all the pastures.

Raising cattle has tradition in the Fulton family, who came to America from County Antrim in Ireland around 1900. They first settled in Kansas. Fulton's father grew up during the years of the Dust Bowl and the Great Depression, but managed to get himself through college and veterinary school and settled in the small town of Loup City in Nebraska. In 1954, he bought the farm. Kevin Fulton went to Kansas State University – K-State for short – and has a bachelor's degree in animal sciences and a master's in exercise physiology. As a student in the track and field team, he started a weight-lifting club at the university. He became a very successful weightlifter and trainer and went on to coach the weightlifting team at MIT, the Massachusetts Institute of Technology. For 27 years, he lifted weights, taking part in 130 competitions, and he is only the second American to lift the 'Dinnie Stone' in Scotland[2].

In 1994, he returned to the farm, and for the next eight years, he farmed conventionally, like his Dad, growing GM corn, soy, and alfalfa. The crop residue was left in the fields over winter, Fulton tilled in spring, and there was no livestock. 'I worked a lot and made no money. Everything went into paying for inputs, seed, fertilizer, herbicides, machinery…. I felt so disillusioned,' he recalls. 'I lost my independence, I was a serf on the land beholden to the likes of Monsanto, John Deere, and Syngenta. I started thinking: how can I regain my independence?' There was no epiphany that made Fulton rethink how he farmed, and the decision to go organic was a gradual process. Nebraska extension agents[3] suggested that a more grass-based system would go a long way towards achieving his goals. 'Suddenly, I realized all the other benefits: quality of life, family life, wild life, soil health, creating an environment that is good for my kids to grow up in, where they can look after chickens, or goats, or set fence posts, where they can have a chance to make a living on the farm later,' he says.

In 2002 Fulton began to revert farmed land back into permanent pasture, with the goal to re-establish native prairie grasses. 'I have a degree in animal science, but I still don't understand how I ignored the relationship between cattle and land, the symbiosis between plants and animals. I scoffed at organic farming,' he says. Those times have long gone. Today, Fulton is often asked to speak at conferences and run workshops. It's important to him to pass on to others

---

2   http://www.thedinniestones.com/Lifters%20Pages/Kevin%20Fulton.html
3   Land Grant Universities were founded in the second half of the 19th century. Their role is to do agricultural research and advise farmers. The latter is done by extension service agents.

what he has learnt over the years as well as the lessons from the setbacks he's suffered. Like Mark Frasier, Fulton is convinced that regenerative and organic agriculture are better for families, for rural communities and for the bank account. 'Farming is a thinking man's game. When farmers brag about the bushels they made or how many acres they farm, I ask them how much money they make, and right now they are losing money on corn. So, I ask them "What are you bragging about?" Unfortunately, your status is still raised by the amount of acres and the yield.' For many, cultivating ever more land with bigger and bigger machinery while chasing record yields still defines a 'good' farmer. And in rural communities everyone knows everyone else's business: to change the way you farm and choose a radically different approach will often be judged with implicit criticism by the neighbors. And any change involves taking a risk, in particular if there is no one around to help and advise. For Kevin Fulton, 2018 is the first year of full organic certification, and so far, he is the only organic farmer in the whole county. Some of his neighbors do believe that he is on the right track, but regarding their own farms, they think they are too old to start something new, says Fulton. Others farm together with their fathers or siblings, who don't think much of regenerative agriculture. Converting to organic also means a three-year transition phase in which the farmer has to adhere to organic practices, but cannot yet sell his produce under an organic label and therefore will not get a better price – in other words: the farm income may drop. Farmers who rent some or all of their land will have to convince the owner to support their decision to convert to organic, because often part of the rent is paid as a percentage of the yield. Still, Fulton believes the hurdles are low compared to the gains that can be achieved in the long run. Over the past few years, the economic situation of farmers in the Midwest has become steadily worse, and experts already believe the situation might get as bad as the 1980s farm crisis. At the time, many farmers were in so much debt that they had no option but to sell their farms. Climate change is one of the reasons for the current crisis; chaotic weather patterns with torrential rainfalls and subsequent flooding have turned sowing and harvesting into a gamble. Seed companies offer corn and soy with different growing cycles: depending on the variety, corn needs between 60 and 100 days from seeding to harvest. Farmers therefore know exactly when the seeds need to go into the ground. During the last few years though, the Midwest has seen extremely heavy rainfall in late April and May, the time in which most seed is drilled. Farmers who farm a lot of land rely on heavy machinery to get the job done, and they need at least one week without rain for the fields to dry off – otherwise the heavy kit just gets stuck in the mud. The industry argues that farmers need to use whatever narrow window the weather gives them and bigger, faster machines will allow them to plant more acres faster. Of course, no mention is made of the fact that bigger machines will also be heavier and need a longer dry period before they can get into the fields. And super-fast planters have a price, with farmers looking at half a million dollars and more. Buying a new planter, tractor or combine is often seen as an 'investment for the future'. 'My son will be able to work with this kit for the next 30 years,' a farmer in Iowa proudly told me. He had just built a machine 'shed' the size of an airplane hangar to store machinery worth several million dollars. In addition to pricy equipment, farmers face rising input costs for GM seed, herbicides, and fertilizers. And commodities, mostly traded through the stock exchange, political factors like the renegotiation of NAFTA, and various tariffs have led to prices dropping further.

Some farmers are therefore forced to sell some land and believe they have to farm the rest as intensively as possible. Kevin Fulton tells the story of a neighbor who has been in financial difficulties for a while whom the banks are now refusing to extend any more credits. Fulton suggested to the farmer to plant cover crops and use the manure from the beef cattle in the small feedlot he is operating as fertilizer to lower his input costs. The neighbor chose a different course of action. Fulton takes a short detour to show us one of the neighbor's fields. He normally avoids taking this route, he says, and I understand why: what had been a row of beautiful, mature trees planted over 100 years ago there now remain only stumps. The neighbor had them cut down to make room for eight additional rows of corn. Financially, that makes no sense at all, says Fulton. And the consequences of this kind of ruthless destruction are already visible: the rain has washed out a deep gully, and one can see the shallow roots of the corn plants sitting in what is not more than a few inches of dark topsoil, while underneath is nothing but sand-colored subsoil. On none of the conventional farms will there be more than three or four inches of topsoil, says Fulton, as that's all that is left from the fertile soil, often several meters thick, that grew over millennia under the prairie grasses. This part of Nebraska sits over the thickest part of the Ogallala Aquiver. 'You could dig a well and find water at 13 feet,' he explains. But the soil now is extremely degraded; modern agriculture has turned those immensely fertile soils into dirt. In Nebraska, dust storms aren't a faint ugly memory from the Dust Bowl decade, but something everyone is familiar with from the traffic news. This year Interstate 80 has been closed several times because of dust clouds. 'Many people have died because they suddenly couldn't see beyond the hood,' says Fulton.

Permanent pastures grazed by cattle are the ideal way to build soil fertility and sequester carbon. But soy and corn growers can improve soil quality, too. When

Kevin Fulton started to use regenerative practices 15 years ago, the soil organic matter was at just 1.5 percent; today it is between 3.5 and 5 percent.

Fulton believes that the reason why the farm today is more profitable than ever before is the fact that he has sold a large chunk of the land to his brother. 'I am on track to make twice as much from 800 acres as I ever did from 2800,' he says. 'Downsizing is a good thing because you can take care of the land. At some point it became more important to have the neighbor's land than to have a neighbor. You just waited for your neighbor to go broke or retire to gobble up their land. I'd rather have more neighbors. We need to stack enterprises and do more with less.' Fulton talks about using land as efficiently as possible and for more than one crop. Instead of planting soy, corn or wheat in spring, harvesting in fall, and leaving the land fallow over winter, a farmer could plant cover crops which could be harvested as fodder or grazed. It's the livestock that adds real profitability because they help to use synergies. A mobile henhouse can be placed on a pasture that has been grazed by cattle. Hogs can break up grassland in a way that otherwise could only be achieved with a plow. Cattle and sheep complement each other in the way they graze. And if they are kept on pastures in alternate years, parasites haven't got much of a chance. They don't survive a whole year without the presence of their preferred hosts. 'I correlate removing animals from the land with people moving off the land. We can reverse the demise of rural communities by bringing back animals,' says Fulton. He'd much rather pay someone a salary for work he or she does on the farm than spend the money on herbicides and fill the pockets of Monsanto or one of the other agrochemical companies in order to save labor costs.

To make the farm viable, Fulton worked with a three-pronged approach: firstly, he increased production, secondly decreased expenses, and lastly, tapped into value added markets and stack enterprises.

Holistic grazing helped to improve productivity. Fulton was able to increase the stock density while simultaneously improving soil fertility and maximizing grass growth. He put in more watering places and more fencing, which allows him to put more cattle on smaller paddocks. In future, he wants to move the animals daily or sometimes multiple times a day. The more successful he is at micromanaging the pastures – from finding the optimal point in time for grazing to choosing the ideal number of animals and leaving them on the paddock for exactly the right amount of time – the better grass regrowth will be. His goal is to rest each pasture for 360 days a year. Because Fulton farms fewer acres, he is able to do the work in less time – he can put down a quarter of a mile of fence (400m) in 15 minutes. To install water troughs and fences was an up-front, one-time investment that will be paid off in a few years. Cover crops extend the grazing time in the year at both ends, and that means he does not have to buy in hay. In fall, he lets the cattle glean the cornfields, and while they are at it, they will also graze the weeds, which Fulton calls 'nature's cover crops'. He then plants turnips which he has the animals graze in December, which brings the number of crops in one field up to three in a single year. Converting to organic has dramatically reduced the input costs. He does not buy any chemical fertilizer or herbicides – neither are allowed in organic agriculture. And because the farm is now certified organic, he can sell his produce at a premium. The combination of regenerative agricultural practices and livestock has increased soil fertility and that has already paid off, literally: in a dry year, his corn yield was twice that of his neighbors who work conventionally. The organic certification creates new marketing opportunities. Because of the lack of infrastructure, the direct marketing of beef remains a challenge, but Fulton is able to sell at least a small amount of meat through an organic food cooperative. The meat he cannot sell through direct marketing might still get a premium as '100% grassfed' and 'local', but often he has to sell some of it into the conventional meat market.

Kevin Fulton's cattle will be out grazing their whole lives. Contrast that with a feed lot. This one is small and not overcrowded.

To have a designated organic meat label would be ideal, but so far, that doesn't exist. And animals that are only kept on grass will reach their slaughter weight at the same time in late summer or fall. Supermarkets, however, need a constant supply throughout the year and take only certain cuts – to set up a functioning supply system under such conditions would take time and money.

Together with daughter Coleen, Fulton would like to build a brand for the farm, which would be a precondition to direct market meat and produce online. And eco-tourism might be a future income stream because there are people who would like to find out what 'fly over country' looks like at ground level.

Fulton's real passion is his Galloway herd, and where holistic grazing management is concerned, he is an absolute perfectionist. But the farm offers so many other possibilities, more than he will ever be able to make use of. Every new project takes time and focus from something else. At the end of 2018, Fulton therefore employed a new (and only) member of staff – a young man who can now make his dream to become a farmer come true. He will spend part of his time working for and with Fulton, but he also has the opportunity to set up his own business on the farm. Fulton shows us the beautiful old farmhouse the young man and his family will be moving into. Then he will have to decide what kind of business he wants to have – growing vegetables, in the field or in a polytunnel, keeping chickens for meat, turkeys or laying hens, pigs, sheep or dairy cows, or maybe using the milk to make cheese… There are farm buildings which could be made into whatever workspace or facility that is needed, and the young farmer will have the use of tools and machinery. Starting a farm business isn't easy, but Fulton will be at hand to give support and advice – he lives in the new farmhouse, up on the hill and just a few minutes away. 'Even after 16 years,' says Fulton, 'farming continues to be a journey with no end point in sight.'

Farming at the margin

The Fresh Seven Café – where one meets in St. Francis, KS

Arranging where to meet with Tim Raile was easy: 9am at the 'Fresh Seven'. It was our second day in St. Francis, a small town in the most northwestern corner of Kansas, and by then we were well acquainted with the 'Fresh Seven Café', 'F7' for short. It's where you get an Americano or a Latte that can hold its own against anything you might buy in a coffee shop in Brooklyn or Berkeley. And the coffee is roasted every day, right on the premises. The 'F7' is where you go to meet people over coffee in the morning, for lunch or for a drink in the evening. Such is the popularity of the 'Fresh Seven' that the local news channel, KSN, ran a short piece[1] on it. We began to understand why a coffee shop is newsworthy a few days before when we made our way from eastern Colorado into Kansas. We had left Mark Frasier's Ranch near Last Chance and drove straight east on Highway 36 for about a hundred miles without passing a single restaurant, coffee shop or fast food outlet. Many of the villages didn't even seem to have a store; occasionally, a flag marked a post office that opened for a few hours a day.

---

1 https://www.ksn.com/news/local/main-street-kansas/main-street-kansas-st-francis-shop-brews-up-coffee-and-conversations_20180306034851450/1011659902

The towering buildings one sees with increasing frequency the further east one gets are grain elevators. Then a road sign informs drivers that they are leaving Colorado, while a second one welcomes them to Kansas, the sunflower state. From the state border, it's another 15 miles to St. Francis and it's easy to keep track of your progress: every mile a dirt road branches off from the Highway, starting with Road 1 and counting.

This is not a toy truck! Grain silos in St. Francis

According to the 2010 census, St. Francis has just over 1,300 inhabitants. There are two motels, and with the support of the town's citizens a third will be reopened soon. St. Francis is the county seat of Cheyenne County, no less, the northwestern-most of the 105 counties in the State of Kansas.

I must admit that we were pretty hungry when we reached St. Francis on a beautiful Monday afternoon in October, and so we made our way to Washington, also known as Main Street. It's where the local school and the chamber of commerce are situated. We walked past the (recently reopened) supermarket, a furniture store, a gun shop, the Farm Bureau, the offices of the local paper, the St. Francis Herald, and, just before we got to the senior center and the towering grain silos next to the railroad tracks, there is the 'Fresh Seven', where we were to meet with Tim Raile two days later.

When Tim and his wife Robyn walk through the door, they are greeted from all sides. Still stacked at the back of the room were some props from a 70s themed party – a few days earlier Tim had celebrated his birthday at the 'Fresh Seven'.

The Railes live in St. Francis; the farm, which was founded in 1902, is situated 13 miles west on Road 2. Tim Raile's family is of German-Russian decent. In the second half of the 19th century, many people from the region around Odessa, on the Black Sea, emigrated to America. A century earlier, their ancestors had left Germany and settled in Russia on the invitation of Catherine the Great, who had promised them freedom of religion, exemption from military service, and permission to keep their language. A hundred years later, the political situation had changed, and the German-Russian community was about to lose its privileges. It was the pending threat of conscription that had more than 100,000 German-Russians emigrate to America. Most were farmers, and they brought to the New World what had been the foundation of their livelihood in the old one: Red Winter Wheat, a variety that still grows very well in western Kansas.

Tim Raile's great-great grandfather arrived by boat in New York in 1885. Ellis Island had not yet been built so he and his family immigrated through Castle Garden on the southern tip of Manhattan. It was the time of railroad expansion across the US, and railroad companies were wooing newly arrived immigrants. Posters and leaflets in ports on both sides of the Atlantic promoted settlement in the Midwest and on the High Plains: there was lots of land available and ready for the taking. The Homestead Act of 1862 made it possible for pretty much anyone to claim uninhabited public land, usually 160 acres, settle there to farm," and file for a deed after five years.

Gottlieb Raile Sr. was two years old when the family went from Castle Garden to Sutton, Nebraska. The Railes spent their first winter in a dug-out. Having seen the remains of a dug-out on Kevin Fulton's farm made it easy to imagine how desperate those first months must have been for immigrants who mostly arrived with little more than what they could carry. The following year, the railroad tracks reached Cheyenne County in Kansas, and in 1887, the town of St. Francis was founded. And that's where the Raile family decided to stay. Tim Raile's grandfather had ten children, eight girls and two boys. He tried to acquire enough land to provide the means of a livelihood for each of them and founded the farm Tim and his son Michael operate today.

While we follow the Railes to the farm in our rental car, it becomes clear why a four-wheel drive in Kansas isn't a luxury but a necessity. It started to rain heavily and just minutes later the dirt road had turned into a treacherous mud track which was as slippery as an icy road. The harvesting crew reached the farm just before the rain set in. Now the heavy machinery sits in the yard and the men are in the barn, waiting for the rain to end. Then it will be time for some maintenance jobs on the big combine, as it will be too wet to harvest. The sunflowers will remain in the field for a while longer.

Tim Raile and the contractor have known each other for a very long time. The work partnership their fathers began is now in the second generation. It's not just good to be able to rely on a tried and tested harvesting team, contractors go where field work is to be done, and it pays to book a slot early. For farmers like Raile, it makes perfect sense to outsource harvesting. Over the years, he has grown a number of different crops, including wheat, corn, and sunflowers for oil or as birdseed. Efficient combines are extremely expensive, an investment that makes sense for a contractor with a dedicated harvesting crew, but not for an individual farmer working with a diverse crop rotation. And the more diverse the operation is the more specialized equipment will be needed. Once farmers use regenerative agriculture practices like intercropping – planting two different crops side by side – it gets really complicated. Ideally, both crops can be harvested at the same time, but to sort them afterwards, more equipment is needed. Working with a contractor who is likely to have state-of-the-art equipment is a good example how a farmer can reduce costs per acre while improving quality and efficiency.

The fall of 2018 was one of the wettest on record in Kansas, but Cheyenne County has its very own micro-climate, says Raile, which he thinks might be due to the elevation, as the farm lies at 4,000ft. Some of the Dust Bowl years were wetter than average. Nevertheless, there were dust storms which blew soil from southwestern Kansas and Oklahoma to the north. Robyn Raile was born in Bird City, a tiny town about 15 miles east of St. Francis. She recalls hearing about the Dust Bowl years as a child, how families tried to seal windows and doors against the dust or found only the top of fence post sticking out of a drift after a dust storm. In parts of Cheyenne County, too, farmers were hit hard by these weather events and got into so much debt that they had to sell off their land, giving others the opportunity to buy cheaply. During the 1930s, many farmers may still have believed that drought years were the exception and the rains would come back. But since the 70s farmers in western Kansas know that conserving soil moisture has to be their top priority; nothing matters more. Initially winter wheat was followed by a summer fallow: the wheat was planted in fall and harvested in the summer of the following year. Then the fields were fallowed until wheat would be sown again in the fall of the next year. 'The summer fallow made this country, farmers got 40 to 50 bushels of wheat,' says Raile.

Weeds were the biggest problem, farmers believed they would use up the sparse water resources in the soil. To keep the fields as 'clean' as possible, they tilled or disked up to eight times during the fallow period; much of the soil remained totally bare for months, exposed to erosion from wind and rain. To plough under wheat stubble was considered to be a good method to add nutrients to the soil. Farming practices changed during the 1980s when glyphosate became available. Farmers ploughed less and controlled weeds by passing with a sprayer four times or more during the summer fallow. And they started leaving the wheat stubble in the fields, which helped catch snow in winter and greatly enhanced soil moisture. While the constant wind often blew the snow right off the ploughed fields, the wheat stubble acted like mini wind breakers, and the ground was more likely to be evenly covered with snow which would melt in spring and increase soil moisture.

Tim Raile didn't always want to be a farmer; initially, he had studied to be an architect. When that had too much to do with art history and not enough with actually building anything, he got really interested in economics and started working with computers which had just started to become more widely used. By 1980, he had gotten married, and when his dad wanted to retire, it seemed like a good time to return to the farm.

In those days, controlling weeds with glyphosate during the fallow period and tilling before sowing winter wheat still was standard practice. Tim Raile added corn to the rotation. In the 1990s, the no-till movement started in the Dakotas as a soil conservation measure. Farmers tried to disturb the soil as little as possible; they stopped deep-plough and kept disking to a minimum. No-till not only reduced soil erosion, it also helped to conserve water. Raile went no-till in the late 1990s and relied on herbicides for weed control instead. He also became an early adopter of GM seeds, and not only used large amounts of glyphosate but any type of herbicide that seemed to do the job, including 2,4-D and Dicamba. 'It was an economical windfall for the farm. Some land went without fallow for eight years. It all depended on how much rain we got. We only fallowed in dry years,' he says. The years until 2011 were excellent on the farm, the yields increased and wheat prices were high. But in 2008, Raile started to notice the spread of herbicide-resistant weeds. 'In 2013, I was at my wits end,' he says. 'We were building a seed bank for weeds.' The worst were kochia, a kind of tumbleweed, and palmer amaranth. Herbicide use increased by 50% within ten years, says Raile, who initially stuck to a wheat-corn rotation but occasionally planted sunflowers because they helped control bindweed and cheat grasses. In 2013, he spent 250,000$ on chemicals to control the weeds. 'I thought this is crazy! It is insane to spend so much on something that isn't working.' Doing the sums was the trigger to go organic. It's a step that makes economic sense and will safeguard the long term survival of the farm, he believes. 'Our farm is not yet fully organic; of the 8,180 acres, only 1,456 are certified organic, 4,320 are in transition, and 2,404 are still farmed conventional,' Raile explains, rattling off the numbers. 'If everything goes according to plan, we will be fully organic by the year 2022.' Not having to use herbicides comes as a relief. He was always very careful applying them, wearing protective gear including gloves and goggles, but 'it was impossible to not occasionally be exposed'. He was 'fairly ok' with using herbicides, except for Gramoxone (paraquat), which is extremely toxic and dangerous when it's accidentally inhaled. Nevertheless, going organic was purely a financial decision: in conventional agriculture input costs are so high and prices so low that the whole model does not make economic sense anymore, says Raile. If there was a weather event like a hailstorm that wiped out the crops he would have to go to his bank for a loan just to keep going. The herbicides were nothing but a cost factor which was only mitigated by what little he got from crop insurance.

Hail damage

Nevertheless, going organic was a lonely decision for Raile and his son. According to government statistics, there are 61,773 farm businesses[2] in Kansas, but

---

2   https://agriculture.ks.gov/about-kda/kansas-agriculture

only 155 are certified organic[3]. In 2013, when Raile had to shell out a quarter of a million dollars for herbicides, the prices for conventional wheat were low, between $4 and $4.50 per bushel. Certified organic wheat (of a good quality) sold for $13 per bushel. In organic agriculture, producing quality is still rewarded, says Raile. In addition, input costs are lower, and Raile firmly believes that in the near future the application of chemical fertilizer will be restricted because of nitrate pollution through farm run-off. That and the shrinking world phosphate resources will, in his opinion, change the way conventional farmers can operate. Still, converting to organic isn't an easy thing to do. 'But I like a challenge,' says Raile. 'I don't like to follow. I wake up in the morning and think: what can I do better.'. He certainly has his work cut out. The family owns about a third of the acreage, while two thirds of the land are leased in sharecropping arrangements. The contracts usually specify that in return for a share of the profit, the owner has to pay part of the fertilizer cost and, in case of irrigated land, a third of the water costs. That means for the farmer that he (or she) will have to discuss with the owner any changes in land management which might potentially lead to a lower profit. During the three-year conversion period to organic, such losses are common. Yields can decrease while the crops cannot yet be sold as organic, which would command a higher price and compensate for slightly lower yields. So farmers like Raile have to convince the landowners first which can be tricky as many of them have a farming background and strong opinions on what should or should not be done in farming. The reasons for absentee landlords to lease their land often date back to the farm crisis of the 1980s. A decade earlier, the then Secretary of Agriculture, Earl Butz, had urged farmers to farm from 'fencepost to fencepost', and to 'get big or get out'. Many farmers invested heavily in land and machinery, but when production reached record level and exports fell, commodity prices collapsed and debt-ridden farmers often had no option but to sell the farm and move to other parts of the country in search of work. Those who didn't have to sell outright leased the land to their neighbors. Tim Raile says the connection to the land runs deep even in the next generation of landowners and he has created a farm website featuring short videos[4] which allows the owners to stay connected. The sums of money involved are huge; Raile and his son manage land worth some $20 million. That in itself is a good reason to convert to organic gradually as there is a learning curve. 'On our organic and transitional acres we primarily grow Hard Red Winter Wheat but have experimented with peas,

---

3   https://www.nrcs.usda.gov/wps/portal/nrcs/detail/ks/programs/financial/eqip/?cid=nrcs142p2_032926
4   https://www.raile.farm/videos/

barley, and oats,' says Raile. During the fallow period he manages weeds through undercutting two to three times – v-shaped knives pulled by a tractor cut off the roots of weeds at a depth of two to four inches. So far, he has not been employing other regenerative practices like cover crops or companion planting; he needs to work out how to do so profitably first. Farming in the High Plains microclimate has its unique challenges, he says. 'With an average annual rainfall of only 14 inches and summer temperatures reaching well over 100°F, growing crops requires close attention to moisture preservation. Our experience from farm trials showed that cover crops were counterproductive to growing crops. Because of our low rainfall and high temperatures, the cover crop depleted the moisture for the cash crop that followed. However, we continue to look for ways to incorporate cover crops that are less moisture consuming and have a shorter growing period. We are in our infancy with organic farming but anxious to learn and try new practices,' he says. 'So for now, on the certified and transitional acres, tillage is our primary practice for controlling weeds. We have made adjustments to our equipment and choice of equipment, and do what we can to minimize soil disturbance. And we maintain the previous crop's residue as mulch to protect the soil from heat and soil erosion in preparation for the planting of the next crop.'

Another huge problem Raile battles with is the lack of infrastructure for certified organic crops, which makes selling them extremely difficult. A conventional farmer delivers his wheat to the elevator of the local coop in St. Francis and 'drives home with a check'. Raile needs to either have a direct contract with a buyer like a food company or he has to sell through a broker. An organic food producer will take 30 days to pay, and the contract usually specifies that the buyer has up to a year for collection. Until then, the grain has to be stored on the farm. And most haulers are not willing to send a truck with a trailer down a dirt road even if the weather is fine. To solve that problem, Raile has bought a small plot of land close to the Highway in St. Francis on which two new grain silos will be built for storage.

Raile's son Michael studied agriculture at K-State and then worked for a company involved in wheat research. The experience he gained and the contacts he made have opened up new opportunities for the farm. A friend and seed breeder has developed a new organic Durum wheat variety. As he was preparing to retire, he gave a bag of those seeds to the Railes for propagation. Durum wheat flour is ideal for making pasta, and winter Durum wheat has huge potential which explains why the research at K-State is funded by industry money. Until now, a lot of Durum wheat was grown in the Dakotas, but as the climate is changing, the number of very hot days in summer is increasing, and that has taken its toll on the wheat production. Kansas still has ideal conditions for growing winter wheat, and organic Durum winter wheat could be a very interesting cash crop: organic hard red wheat sells for $12 to $13 a bushel, organic Durum for $18 a bushel. Raile has only had one season growing organic winter Durum, but so far the results have been good; yields and protein content were comparable to those of red wheat. When he still farmed conventionally, increasing the yield was the top priority. 'My priorities now are first of all quality as in test weight[5], then protein content, and thirdly yield,' he says. Talking to Tim Raile underlines that an organic farmer will fare better with a bit of pioneer mentality. There are no recipes and no easy answers. Farmers like Raile need to find their own paths and their own solutions for each and every problem on the farm. 'Most conventional farmers still have the attitude "this is what I know, my father did it that way, it worked for him, it's good enough for me",' says Raile. 'Back in the day, there was always a fix. Now we have to be proactive and choose the right seeds, they need to be fungus- and drought-resistant, and they need to create a good canopy fast to shade out the weeds. We now seed wheat closer, we switched from a 12in drill to a 10in drill to supress weed growth.' For the same reason, Raile will no longer grow sunflowers or corn; the space between the plants is wider, which increases weed pressure. Since he started farming organically, Raile has developed a new and better understanding of soil quality, but in a region with so little rain, compromises have to be struck, and while soil quality is important to him, it isn't the top priority. Since he reduced tillage to a minimum, the soil organic content has risen from 1.5% to 2%-2.5%, and Raile is optimistic that organic practices will further improve soil quality. Hopefully, that will increase the water-holding capacity of the topsoil. Raile says there is often quite a bit of subsoil moisture, but the top few inches are very dry.

For now it's one step at the time. Raile wants to complete the conversion to organic and then decide what other changes to introduce. Which doesn't stop him thinking about the options. Cow peas are an excellent source of protein and could be sold to poultry farms and pet food manufacturers. It's a short season crop that is sown in spring and harvested in July. Chick peas and quinoa may be other possibilities – what Raile is looking for is drought- and heat-tolerant crops for which there is a market and a profit margin to be had. What will never work is crops with a growing season stretching into July and August – it's just too hot for that in western Kansas, he says.

---

5   The test weight is the quantity (measured in lb) of wheat that can be contained in a standard volume, in this case a bushel. The higher the test weight the better the quality.

Raile is also working with two companies developing algae-based bio-stimulants, which in his opinion could be very beneficial and worth testing on the farm.

As for the future, Robyn and Tim have three grandchildren, two boys and a girl, and the Railes want them all to have the choice to remain on the farm and have a future in the business. Farming and farm management are just one option. Like Kevin Fulton in Nebraska, Raile too wants to improve marketing opportunities and develop a specific farm brand. The farm could also offer services, from brokering organic grains to creating and managing storage capacities. And, as with Fulton, Raile would like to reduce the acreage of the farm; at some point, he only wants to farm the land he owns outright. It's about making more profit from fewer acres by farming them better. 'The only way we could go on as conventional farmers was to add acres in order to stay where we were or grow the business slightly. That's not sustainable,' he explains. 'We need to optimize what we do, and we need the time to do it. You cannot manage so much land in the best possible way because you cannot pay the same attention to detail.' Organic agriculture is labor-intensive, and some of it is unskilled, manual work, but there are increasingly more jobs for highly skilled workers and professionals. This year, Raile will hire an extra person, a young farmer with a degree in agriculture from an agricultural college or a university. Long-term, Raile can imagine producing a range of products on the farm, from organic pasta made from Durum winter wheat to protein powder from organic peas.

Tim Raile was the second farmer in St. Francis to go organic. The first one was Robert Klie. His 2,100 acre farm is a little to the west, on Road 1. It's been certified organic since 2005. When he took the decision to transition to organic, the neighbors didn't hold back: 'It was: that crazy Klie, he doesn't know what he is doing, but I showed them,' says Bob Klie and laughs. The Klies own most of their land outright; only 300 acres are leased. They grow wheat, corn, oats, peas, and clover for feed: 450 acres are permanent pasture for 36 cow-calf pairs. Apart from beef cattle, the Klies raise between 15 and 75 pigs for slaughter every year. Certified organic pork and beef should make for a good profit if it could be sold as such. But that's impossible given the lack of infrastructure. For slaughter and processing, Klie has to drive the animals either 2.5 hours to Beaver City, Nebraska or to the nearest slaughter facility in Kansas, which is in McPherson, a five-hour drive to the east. Neither facility is certified organic, and that means though the animals have been raised organically, they cannot be sold as such. This type of infrastructure problem doesn't just exist for organic farmers but for producers of conventional meat, too, at least if they decide not to play by the rules of the big meat processors.

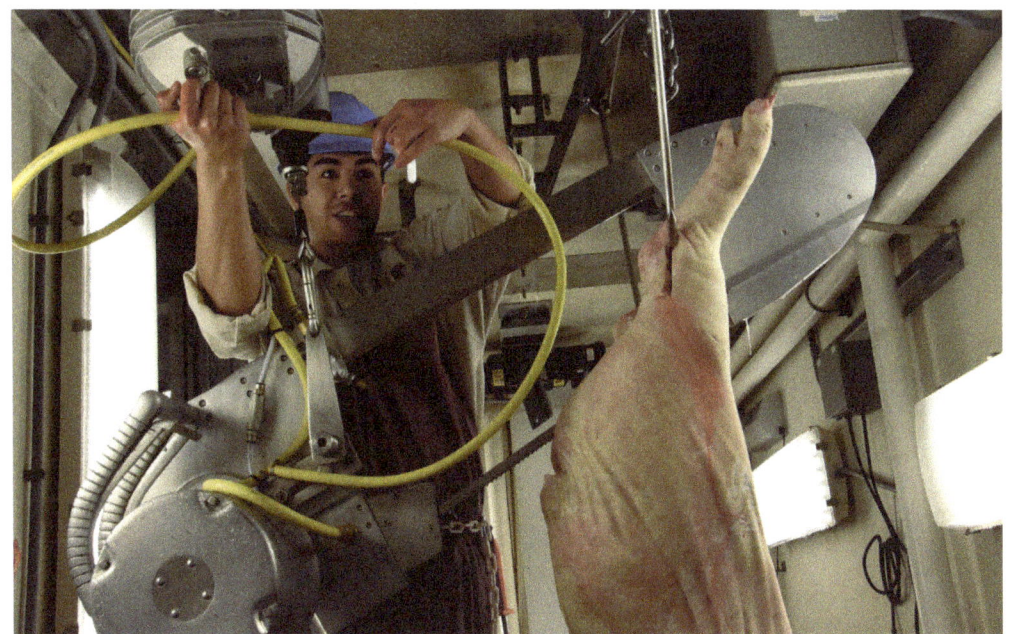
Inside the mobile slaughter unit on the Callicrate Ranch

Just outside of St. Francis, a huge sign indicates the dirt road that leads to Mike Callicrate's ranch. He has a small Wagyu beef herd, he raises pigs, and he works with about ten farming operations in the region who supply him with calves for finishing. Once the animals have reached their target weight, they are slaughtered on-farm in a mobile slaughter unit and then transported 200 miles to Callicrate's processing facility in Colorado Springs. Different cuts of meat, sausages and hamburger patties can be ordered online and are available at a store next to the processing unit. A lot of the meat goes directly to restaurants and diners, the 'Fresh Seven Café' in St. Francis being among them.

Mike Callicrate didn't exactly dream of setting up a farm-to-table operation, but in the end, doing so was the only way to save the ranch.

He grew up on a ranch in Colorado, and after college he worked in a number of different jobs – from marketing to being a professional bull rider – until he married and moved to St. Francis, taking on the management of his wife's family farm.

In the late 1980s, while the consolidation in the US meat market accelerated, more and more family-run slaughter facilities and processors were forced out of business or were bought up. The few remaining big meatpacking companies started to flex their muscles. Callicrate says that when he started ranching he had the choice to sell his animals to about 20 different packing companies in Texas,

Farming with Benefits

Kansas, Nebraska and Colorado, but by the 1990s only a few conglomerates were left and the cattlemen felt the effect: in the 1970s 70% of the profit from meat sales went to producers; today, he says, it's 30%. After he joined a class action lawsuit against the food company and meat processor Tyson (formerly IBP) with charges of price fixing and acting as a monopoly against the interests of producers and consumers, Callicrate found that he was unable to sell his cattle at all.

From the air: center pivot irrigation in Colorado. Crops wouldn't grow here without it. The sprinkler 'arms' can be up to half a mile long.

He wasn't the only one; even ranchers not involved in the lawsuit struggled, says Laura Krebsbach[6]. She came up with the idea of a mobile slaughter unit in 2004, and together with a small group of co-creators holds patents on the design of the unit. Krebsbach is a trained paralegal, and in the early 2000s, she was based in Nebraska, working for an advocacy group and the 'Renewable Harvest Project'. 'The big meat processors flooded the market with cheap meat. Farmers either went with these big companies, or they had no access to the markets. The USDA-run facilities closed one by one, so for farmers there was no way to get animals to slaughter'. Krebsbach's goal was to give farmers market access so that they

---

6   Laura Krebsbach now lives in Arizona. I talked to her before the trip and again at length a few days after we had visited Mike Callicrate.

could directly market their meat. A mobile slaughter unit seemed like an obvious solution. To work out what specifications such a unit needed to have in order to be efficient and get the necessary USDA approval, she assembled a small group of experts and farmers; one of them being Mike Callicrate.

The HACCP (Hazard Analysis and Critical Control Point) system establishes the ground rules: the unit needs to be high enough to hang a beef carcass, the workspace has to progress from dirty, to clean, to cleaner, and finally to the freezer. There needs to be a way to present the animal's head and organs to the USDA veterinarian for inspection. Hot water (82°C, or 180°F) needs to be available at all times, and there has to be a backup generator, should the outside, on-farm power supply fail. Krebsbach managed to source six custom-built so-called 'drop deck or drop belly' reefers from a meat processor that had switched to a new system and no longer needed them. The reefers had been used to transport beef halves from small slaughter facilities to bigger processing plants. They therefore had lowered axles to provide the required extra height, a reinforced hull, and rails on which the carcasses could be moved. A small construction company in Nebraska transformed the reefers into moving abattoirs: a control unit for electricity, temperature, hot water supply and appliances needed to be fitted, the 'dirty to clean' work stations had to be designed, and additional rails had to be installed for shifting the carcasses into the reefer.

The mobile slaughter unit can be operational pretty much anywhere with access to a water and electricity supply, but a few things need to be in place around the unit: a corral has to be set up so that the animals can be delivered the day before and have space to rest. And there has to be a chute that serves as a kill box. The whole unit can be adapted to the slaughter of cattle, pigs or sheep and operated by just two skilled slaughtermen. Once the animal has been stunned and killed, a front loader lifts the carcass so that it can be attached to the movable rail that extends from the slaughter unit. Then the head is removed and presented for inspection to the USDA veterinarian. Once the animal is bled out, it is transferred into the unit, the doors are closed and a slaughterman can start processing the carcass. The cattle hide is removed inside the trailer, pigs will be scaled and the bristles removed outside.

Cattle on the Klie farm

On Mike Callicrate's farm, the unit is in use five days a week, with cattle and pigs being slaughtered on alternate days. We visit on a Tuesday, and that means it's a pig day. With the ranch in north-western Kansas and the processing facility in Colorado Springs, the mobile slaughter unit is the link that connects both operations and makes them into a seamlessly functioning, economically viable system.

Callicrate has set up a standard warehouse with a docking station for the reefer. While the work could be done outside, having protection from the weather makes things a lot easier. The kill box is stationed inside, and the permanently installed rails and chain winches for moving the carcasses are much more efficient than using a front loader. With four people working outside the reefer in the warehouse and one inside, 20 pigs or 10 head of cattle can be slaughtered in a day. The mobile unit is spacious enough to allow us to be inside and watch what's going on without being in anyone's way.

The intestines and organs are pushed through a flap and land in two separate boxes for the USDA inspector to check. Mike assures me that a USDA official is present until the last carcass has been inspected and deemed fit to enter the food chain. When I later ask Laura Krebsbach about this she says that the USDA is required by law to provide an inspector for eight hours a day without extra cost. 'Initially, the veterinarians were really unhappy,' says Laura Krebsbach. 'They considered it an expensive waste of time. But it didn't take long until they started to volunteer for shifts at the mobile units, saying: we are actually doing our job here! In a big unit, they are usually just on the kill floor because they are only authorized to intervene when they see inhumane killing.'

Without the two mobile slaughter units, he wouldn't have been able to keep the ranch and rebuild his business, says Callicrate. When the scheme got off the ground in 2004, it cost just $130,000 to refurbish one unit. Six were built, and all are still in operation. The reason the units could be made so cheaply was that the company at the time wanted to get rid of the 'drop belly' reefers and was happy to sell them off for just $ 5,000. There is a huge demand for more such units: if they had access to a certified organic mobile slaughter unit, the Klies could sell the meat from their organically raised cattle as organic, with a much higher margin. Farmers like Kevin Fulton in Nebraska would be better able to market their organic meat; a remote location wouldn't pose much of a problem anymore. So why aren't there such units operating across the country? Krebsbach has tried in vain to source more reefers but 'they seem to have vanished from the face of the Earth'. According to the company, they were all sold to Mexico, says Krebsbach, except that she was unable to trace a single one. And new reefers aren't an option either: the only company that was prepared to build new units at a cost of $100,000 each withdrew the offer when Krebsbach wanted to order ten. The million dollar order was suddenly considered too small and not worthwhile for the company. Laura Krebsbach continues to look for ways to provide ranchers with direct market access. Concentration and vertical integration in the meat sector continues, she says, and the more control processors have over ranchers, farmers and finishing units, the better it is for their bottom line.

But back to Bob and Joanne Klie's farm. The Klies took the decision to go organic in 1999, practically at the same time their neighbor Tim Raile opted for going down the industrial farming route and growing glyphosate-resistant GE crops. For Klie and for Raile, the decisive factors were financial considerations. Raile went for higher yields, Klie for a higher margin. At the time, conventional corn and wheat sold for $2 a bushel, organic corn for $8, organic wheat for $6. Klie says he never regretted his decision to go organic. Yes, the yields went down a little, but with the better price he gets for his certified organic produce and low input costs he's been able to make a profit year on year. While conventional farmers had to spend more and more on agro chemicals, Klie only had to buy seed. He ploughs the wheat stubble under; tilling remains his preferred way to control weeds. Five or six tractor passes in a season on a single field are not unusual for him. He doesn't grow cover crops because he thinks there isn't enough moisture. And yet he says he was surprised how fast the soil quality on the farm improved once he went organic. The water infiltration rate is about two inches an hour. On neighboring farms, it is a quarter inch at best, he says, and the rainwater runs off brown, taking topsoil with it.

For his wife Joanne, better soil quality is just a positive side effect. She has worked for years as an assistant physiotherapist at the hospital in St. Francis. She still remembers the lecture she attended about the use of agro-chemicals and the increasing number of cancer diagnoses in rural areas. Throughout her career she had met cancer patients almost on a daily basis, she says, and that's why she believes organic agriculture is so important. To her, eating organic fruit, vegetables and grains is a precondition for staying healthy.

In all our conversations with farmers, the cancer topic came up at some point: a family member, a neighbor, a friend, or someone's child had been diagnosed with the disease, was in treatment, had died... One of the farmers called to postpone our visit to the following day – the wife of his neighbor, a woman in her early thirties, had just died of cancer, and he would attend the funeral.

Joanne Klie takes pride in her vegetable garden and shares what she grows with friends and neighbors. But on June 19th 2018, a hailstorm moved across Kansas with 90mph winds and hail the size of golf balls. Joanne's vegetable patch and 1,000 acres of oats and wheat were completely hailed out. After the storm the Klies first went to check on the cattle, but the animals had found some shelter in a lower lying area of the farm and were alright except for some bruises. The damage to the farmhouse was quite severe, and at the time of our visit one of the back windows was still boarded up. In the yard, the Klies show us their trailer which looks as if someone had attacked it with a hammer. But the 'High Plains Food Coop' logo is still clearly legible. As we walk to the farm building where the Klies store meat in several freezers and keep a small grain mill, they tell us about High Plains. In 2008, a group of farmers founded the cooperative to direct-market their produce. Today, there are 50 farmers and producers who sell through the coop and run it on a volunteer basis. The product range includes organic and conventional produce; any use of GMOs has to be clearly labeled 'and is frowned upon'. Joanne Klie works between three and eight hours a week for the coop: she maintains the website, responds to customer requests and deals with orders. Until the second Thursday of the month, orders can be placed, and the delivery date is a week later. On delivery day, the Klies have a very early start. The collection point in St. Francis is a storeroom with access to a kitchen on a private property. There, the orders are sorted, packed, and loaded onto the trailer. By 7.30am, everything should be ready to go – the round trip to Denver and back takes roughly six hours.

Most of the coop's customers place relatively small orders; only ten restaurants and coffee shops buy regularly, mostly burger patties, eggs, and flour. For the Klies, even small and irregular orders are worth their while. Before the coop was founded, they had to sell their cattle either locally or through an auction in Denver. They seldom made more than $400 or $450 per head of cattle. Since they sell meat, sausages and hamburger patties through the coop, the profit per animal has been around $7,500 – minus the cost of $1,000 for slaughtering and processing.

Jeter Isley shows us what hopefully will grow into a hedge

With grains, the profit margin is equally high. The Klies sell most of their grains through a broker. Coop customers need small quantities, but the Klies offer a variety of specialty flours like Emmer, Einkorn, Turkey Red Wheat, and Triticale. Investing in a grain mill has paid off handsomely: a 50lb bag of Triticale sells wholesale for $3 to $4. Coop customers buy a 2.5lb bag of freshly milled Triticale flour for $5 to $6. The flour is delivered together with one of the pretty cloth bags Joanna Klie makes. Customers love those little bags, she says, and as she uses cutoffs, each one is different. With her skills and a sewing machine, it doesn't take her long to make the bags and keep her customers happy and feeling a little special.

With a turnover of $40,000 annually, the Klies are among the coop's bigger suppliers, and Bob and Joanne agree that it would be hard to stay profitable without High Plains Food Coop.

But like other coops, High Plains, too, is in crisis. The Valley Food Coop in Colorado had to close, as had the Oklahoma Coop, which initially served as a model for setting up High Plains. At present, the board is in talks about how to future-proof the cooperative, says Joanne Klie, who has been the coop's (unpaid) treasurer for a long time. But to discuss the problems High Plains faces and the changes that could be made, she suggests we talk to the coop's president, Jeter Isley.

That's not a problem as we are scheduled to visit him anyway.

Jeter and Nina Isley run a certified organic farm outside Bird City, east of St. Francis. We've already sampled some of the 'Y Knot Farm' produce; the Isleys supply salads and tomatoes for the 'Fresh Seven' burgers. There are two hoop houses for growing tomatoes, cucumbers, peppers, eggplants, and herbs; other vegetables are grown outdoors. Add to that fruit trees, berries, organic wheat, and meat from their Galloway herd.

On the day of our visit, an icy wind blows steadily from the north, and we retreat to the warmth of the old farmhouse to discuss the problems the High Plains Cooperative faces, over a cup of coffee. 'We need to up our game or fold,' says Isley, who wants the coop to be run more professionally. The annual turnover of High Plains is just $200,000, and the Klies and 'Y Knot' are among its biggest suppliers. Isley believes it's untenable for the farmer-members to continue running the cooperative on a volunteer basis. Farmers are busy as it is, and it is just not feasible that they are burdened with everything, from admin to making the deliveries, without getting paid. Self-exploitation isn't a sustainable business model and Isley notes that 'the founding members are getting older'. He believes the coop has to grow in order to survive. The deliveries have to go from a monthly to a weekly basis, the customer base has to be broadened, and he would like to hire someone who runs the day-to-day business in a professional manner: 'Farmers know how to grow stuff or raise cattle; marketing needs a totally different set of skills. We need a model where everyone does what they do best.' The number of suppliers has to grow, too. Ten to twelve farmers should produce enough to sell produce worth between $10,000 and $20,000 annually through the cooperative. And the range needs to be extended, possibly by cross sourcing from other coops. The Tap Root Coop, a Colorado-based cooperative, could be a good match; some produce could be sourced from them if they in turn were to carry High Plains products. Customers should be able to buy what they need when they need it. Isley hopes the board will agree to his suggestions. The changes would pave the way for High Plains to reach out to customers from Colorado to eastern Kansas. The collection points could of course serve as pick-up points for customer orders, too. It would open up opportunities for people in sparsely populated, rural areas to buy, for the first time, high-quality produce from local farmers at an affordable price. Right now, large parts of the Midwest and the High Plains are food deserts where people have to buy food at gas stations or small convenience stores because the nearest supermarket is a long drive away. The supermarket in St. Francis remained closed for several years and was only reopened recently, after intense lobbying efforts by citizens. On sale are a lot of ready meals, but there is a cooler with fresh dairy and meat products and a selection of fruit and vegetables, most of it from California or Mexico. Supermarkets usually get deliveries from central distribution centers; they are often unable to make individual contracts with local farmers like 'Y Knot Farm'.

At present, Jeter and Nina Isley sell mainly meat through the High Plains Coop. There is a lot of demand for certified organic fruit and vegetables, but most varieties can't be stored long enough. 'It is impossible to market fresh produce on a once a month basis,' says Isley, 'for fruit and vegetables deliveries have to be weekly.'

In many ways 'Y Knot Farm' is an unusual business. Nina Isley, who grew up on a farm in Nebraska, was the driving force behind going organic, says Jeter Isley[7]. 'She's attuned to the earth, she cares for the health of the soil and the health of plants. Food quality really matters to her and she hates to see the neighbours applying chemicals, because she can smell them. Her mantra in farming is: What comes off their land should be good.' Jeter Isley shares that world view, and he is convinced that a business based on these principles can be financially successful, too. 2018 will probably be the first profitable year for the 'Y Knot Farm'. Until then, it will have been a long and difficult journey.

Jeter Isely was born in New Jersey in 1954. Both his parents died in quick succession, and by the age of three, he was living with French-Canadian relatives. He returned to the US as a teenager, first to go to a boarding school in Connecticut, then to study history at a college in upstate New York. After graduation, he spent the next three years travelling from Europe to Asia and then on to Australia and New Zealand, working along the way. During his time at a sheep station he made observations that are proving valuable for decision-making 'Y Knot Farm' today.

---

[7] On the day of our visit Nina had to attend a meeting, which unfortunately meant we were unable to talk with her directly.

After his return to the States, he worked for three yeas as a cowboy in Montana in fall, winter, and spring, and as a teacher in summer. His next career move took him to Colorado, where he and a group of friends started gold mining. Narrow vein mining had been shut down in 1942 to free men for the war effort. After the war ended, most of the mines never reopened, but by the 1980s there were new ways of gold extraction with chemicals that suggested the mines could become profitable again. 'No one in our group had the skills to make it work financially,' says Isley, who decided to go back to university. He got an MBA in finance from Cornell and joined the chemical conglomerate FMC in 1989. He was a director of finance for 15 years, until in 2006, he and Nina decided that they'd much rather spend part of the year sailing and otherwise swap life in Chicago for one in a rural area, preferably like the one where Nina was raised. The 1,100 acre farm they finally bought is just 30 miles south from where Nina grew up. Before the Isleys could even think about farming, they first had to make the old farmhouse habitable once again. The land was rented out for the first two years, but in fall of 2007 chemicals were applied for the last time. Nina started growing vegetables straightaway and did all the required courses for handling produce and selling it into the food chain. But before taking any major decisions for the future of the farm, they wanted to get to know the land, the soil, the climate, the weather, and the seasons. Organic certification, sustainability, soil health, and diversity were as much a goal as building a financially viable business. Jeter Isley, too, understood quickly that none of these aims would be achievable without sufficient water. How much water is needed? How much moisture is lost? How much water can be saved? The answers to these questions still influence every decision that is taken on the farm. 'Y Knot has three legs to stand on,' explains Isley, 'organic wheat, beef cattle and growing vegetables.' 700 acres of the farm are permanent grassland, while the rest is used to grow wheat. The constant wind is enough to dry out the land. The cash crop, wheat, grows over winter, and planting summer cover crops would be ideal. 'I am a fanatical believer in cover crops, but we need the moisture,' says Isley, confirming what Tim Raile and Bob Klie have already told us: cover crops use moisture that is needed for the wheat to germinate in fall.

While most conventional farmers try to farm large, contiguous areas, Isley reduced the individual size of his fields to minimize soil erosion and water loss through evaporation. 'And you want to make sure you leave as much organic matter as possible and add manure and compost. Sometimes you need to invest in water to preserve it,' says Isley. Since 2012, he has planted more than 5,000 trees and shrubs, and he has plans to continue for another few years. The trees are planted in rows of three in a north – south or east – west direction. That way, they will grow into a shelterbelt against the fierce winds from the north and west. The trees need to be irrigated during the summer months for six to eight years, otherwise they won't survive. 14 to 18 inches of rain may be the average for Cheyenne County, but in some years there is a lot less. Newly planted trees need to develop long tap roots to be able to deal with an extended summer drought, and until then they need to be watered. Isley shows us some of the planted hedges; the soil is covered with plastic sheets to keep the weeds down – weeds, too, compete for water. Along the side of the trees, close to the stems, runs a pipe for drip irrigation. Isley turned to the specialists at the NRCS, the National Research Conservation Service of the USDA, to work out a planting plan and choose suitable varieties. The assumption is that the height of the trees times twenty will be the area of the field that is protected from wind and thus from moisture loss and soil erosion. Isley learnt about the benefits of windbreakers in New Zealand. The summer season on the South Island is hot and dry. On the sheep station Isley worked on there were wide, mature hedges, and he noticed that the grass on either side remained green for about three weeks longer than on neighboring farms without windbreakers. For a farmer to have an additional three weeks in which he can buy less feed or even none at all can add substantially to the profit margin. In some years, it can make the difference between survival and going bust. On 'Y Knot Farm', the trees take up 3% of the available agricultural land, but they also boost the wheat yield by 20-25%. Isley has planted more than 30 different species, many of them fruit trees and berries, elder, buffalo and mulberries, which will provide an additional income. The wood of locust trees will eventually make some very hardy fence posts. Other trees are just beautiful. Isley planted cottonwood for the sound the leaves make, forsythia for the deep yellow of the flowers, and fragrant sumac for the birds. Osage is extremely hardy and out of the 200 he planted, he lost only four. There are also burr oaks, ponderosa pines, blue spruce, and grey and red dog woods, to name just a few. The NRCS only had experience with Rocky Mountain juniper and red cedar, but they were happy to help source other varieties, though, as Isley notes, 'they certainly took their sweet time'. Planting trees as windbreakers isn't a new idea. During the Dust Bowl of the 1930s, the FDR government supported the planting of more than 200 million trees throughout the Great Plains to prevent soil erosion and create a barrier against the dust storms. Without irrigation, many of the trees died, and by now farmers have ripped out most of the remaining shelterbelts to create larger fields that are easier to cultivate. It's not surprising that the NRCS extension agents were eager to help a farmer who wanted to actually plant trees.

Isley hopes that with organic farming practices like composting and mulching the soil organic matter will go up considerably. That should in turn increase the water infiltration rate and the water holding capacity of the soil. Together with the windbreakers, there might soon be enough soil moisture to plant summer cover crops – which would improve the soil quality even further.

Still, right now planting cover crops remains a big gamble. Isley has made several attempts, seeding peas and oats in March. Oats provide shade for the peas, but if there isn't enough rain in May everything just wilts. Cover crops could set a positive, upward spiral into motion: peas don't produce just biomass but also nitrate, which is a readily available fertilizer for the next crop. Isley estimates that peas can provide roughly 75-150lb. of nitrogen per acre. But seed costs are high, cover crops are a lot of work, and if there isn't enough moisture, they are a total loss. Between June and September nothing grows, says Isley, but if cover crops can be undercut at the end of May they will form a thick layer of mulch that protects the soil underneath.

With wheat yields increasing by up to 25%, the financial benefit of the hedges can be calculated directly. Add to that the berries which the Isleys could harvest and sell for the first time this year, making the fruit trees and berries become really profitable if and when the High Plains Food Coop switches to weekly deliveries. Berries have to be harvested as soon as they are ripe, and cannot be stored. The same goes for the peach trees Nina Isley has planted next to the hoop houses. Hedges and trees provide secondary benefits as well: the Galloways and their calves can find shade and protection from the wind. Only for the hoop houses did the Isleys decide not to rely solely on the hedges. They constructed a sturdy windbreaker from old electricity poles and pallets. The makeshift wall already proved its worth when it withstood an 80mph storm, and the hoop houses remained undamaged.

'Y Knot Farm' would already be profitable had it not been for a series of unexpected setbacks. In 2009 the farm was hit by as thunderstorm of such ferocity that every tree and bush was completely stripped of leaves. 'They were totally bare,' recalls Isley with a shrug. That's how nature renews itself.' In June of 2018, he and Nina had stood outside their house watching the hailstorm that devastated Bob and Joanne Klie's farm. That it moved south and didn't hit 'Y Knot' was pure chance. In 2017, their farm was on track to make a profit until they lost seven cows in one night to a mountain lion. 'He killed the cows and our profit'. Because of predators, Isley has given up raising chickens. One racoon killed three quarters of the entire flock in a single night. Others were not as vicious, but neither nightlights nor leaving the radio on deterred them. Isley even tried different stations, but rock music proved to be as useless as NPR[8]. The following year, 2018, they invested $3800 in a new bull that was then attacked and killed by another bull. 'We have no idea what happened,' Isley says, 'but luckily we will make a small profit this year, nevertheless.'

Potentially there are lots of opportunities on the farm to increase profits. Processing fruit and vegetables that are over-ripe, too small or not perfectly shaped into jams, chutneys or relishes, canning fruit, pickling, or making soups and stock using the bones from their own animals as well as cheap cuts that don't sell – all could be done if it weren't for health and safety regulations. 'These rules are geared towards big companies,' says Isley, 'but in a farmhouse kitchen you can't comply with rules that apply to a large commercial kitchen as we have neither the space nor the appliances.' And then there would be the paperwork. 'Big companies have staff members who don't do anything else but fill out the right forms,' he says. In many cases, samples have to be taken from every batch and stored so that an ingredient can be traced back should a problem emerge. Being so close to the Colorado and Nebraska border, would the Isley's want to sell across state lines? 'Forget it,' says Isley, 'the amount of regulation gets even worse. But it is really sad. Especially in an area like ours we really need small businesses that produce high quality products.' Such enterprises would create jobs for people with very different skill sets and qualifications and with a chance to make a living, and they could stay or return to small, rural communities. Isley believes that there are a growing number of people, young families in particular, who are tired of the constraints of city life or simply want to return to where they grew up. At present, there is a labor shortage in Cheyenne County. In summer, Isley tried without success to find workers to help with the fruit harvest, offering to pay $13 an hour (the minimum wage in Kansas is $7,25).

For now, selling through the High Plains Food Coop is not just the best option, it is pretty much the only one. That's why so many hopes ride on hiring a professional to expand the delivery service and customer base. What's also needed is a better understanding among customers that there is a connection between good food, health and the environment. In south-east Kansas, two farmers have set out to promote just that.

---

8 The tech podcast 'Reply all podcast' was at least in one instance successfully used to keep peach tree in the Catskills safe from a bear. https://www.businessinsider.com/tech-ceo-uses-podcasts-to-defend-his-peach-tree-from-hungry-bear-2018-8?r=US&IR=T We heard the story on https://www.wnycstudios.org/story/breaking-bad-news-bears

Farming with Benefits

Farm Futures

Inspecting the fresh pasture on Gail Fuller's farm

Emporia is a small town in Kansas, about half way between Kansas City and Wichita. It is home to Emporia State University, and it is the start and end point of 'Dirty Kanza', the annual 200-mile gravel bike race which attracts participants from around the globe – their signatures on the wall of the coffee shop on Main Street testify to that. At night, you hear the low rhythm and unmistakable sound of the train horns announcing another mile-long freight train slowly rolling through the town. Emporia is an important railhead, in particular for corn, soy, and wheat. In regenerative agriculture circles, Emporia is known for another reason: situated at the edge of town is the farm of Gail Fuller. As early as 1993, Fuller retired his plow and went no-till. Since then, soil quality, building humus, and looking after the community of soil organisms have been more important to him than high yields. For Fuller to become a regenerative agriculture pioneer and by now a bit of a celebrity was something he never aimed for nor considered likely. 'If there's a mistake to be made, I've done it twice,' he says.

We're waiting for Fuller in the shade of a beautiful old tree in the farmyard. A few chickens scratch between grass and gravel, while two large sows lying in a fenced-off paddock watch us with interest. A few minutes later, a pickup pulls into the yard and Gail Fuller gets out. Tall, in t-shirt and jeans, he has something of a lanky teenager about him. He is friendly and open, focusing intensely, with his eyes not just observing but 'reading' details and minute changes. Only later do I realize that within minutes we were in a deep and very personal (reciprocal) conversation: aging parents, divorce, the sudden death of his brother, dyslexia… He finds dealing with figures and written communication, in particular via email and on a computer screen, difficult. Fuller says he perceives and understands the world visually, through images and observation. Even as a child, nature and the environment were very important to him. He was a teenager at the tail-end of the 60s protest movement in the US, a time when processed food produced by Big Food became deeply unpopular and many in the beatnik generation returned to the land and started to grow organic food, for themselves and others.

Fuller recognized early that soil erosion had become a huge problem. In 1993, the Missouri burst its banks, and in south eastern Kansas, large areas were flooded. Fuller recalls how the water submerged part of a field he had just plowed. He reckons he lost eight inches of top soil that year, washed off by the floods. He decided to never let that happen again and went no-till. Around that time, he started to experiment with planting cover crops, though with poor results. 'I just didn't treat them like a real crop. I was young, arrogant, and not willing to ask for help,' he says now. He brought wheat back into the soy-corn rotation. Wheat stubble protects the ground from rain and wind erosion, and in winter, it catches the snow. Farms in arid regions of the US need a thick, regular snow cover on their fields over winter. As the snow melts in spring, the water will infiltrate the soil, and the additional moisture can make the difference between seeds germinating and a crop growing, or not.

In 2002, Fuller once again began sowing cover crops, and since then the soil quality has markedly improved, he says.

Fuller talks while we sit in his pickup on our way to the field on which he started his cover crop experiment. It was the worst field on the farm, where his father used to run a small feedlot. The soil was very compacted and in bad shape with the subsoil visible, knobs of clay sticking up everywhere and the humus layer almost completely gone. The phosphate content was extremely high and initially nothing but the most phosphate-loving weeds grew: dandelions, bind weed, legumes, alfalfa, and ragweed. Like Klaas Martens, Fuller values them as indicator plants that tell him about the imbalance in the soil. Over ten years, from 2002 to 2012, that field became his 'sand box' in which he just tried things out. He had animals under- and over-graze parts, he worked with and without glyphosate and grew stuff that nobody in Kansas had ever before seen in a pasture, including hemlock – he always hoped no cow would die. Within ten years, the soil organic matter grew from two to seven percent. The field day was in 2012, when soil scientists found that there now was a layer of topsoil between two and four inches deep. The knobs of subsoil had vanished completely. Fuller gets the spade from the back of the pickup and takes a soil sample – it's dark, crumbly, and full of worm channels.

Gail Fuller in one of his experimental fields

The 'sand box' field turned out to be a success story, the rest of the farm less so. And that has to do with the Farm Bill, the law that governs US agriculture and food policy. The Farm Bill covers a huge number of issues, from SNAP[1] – which provides food aid for low income families – to disaster assistance for farmers who have suffered losses through catastrophic weather events such as flooding. But there are also so-called 'safety net and price support programs', which are basically an income loss insurance, mainly geared towards corn and soy growers. Farmers who want to benefit from such programs have to adhere to a number of strict rules which, from a practical farming perspective, can be counterproductive: regenerative agriculture does not fit into the system. A farmer planting cover crops

---

1 SNAP stands for the Supplemental Nutrition Assistance Program.

in fall to prevent soil erosion in winter has to 'terminate' them with glyphosate before a certain date. However, to prevent drift and harm to neighboring crops, glyphosate can only be used when there is little or no wind. The previous year, one of Fuller's neighbors, a tomato grower, had suffered drift damage, so Fuller went ahead with seeding corn and soy, but delayed the glyphosate application until the weather conditions were right. For the insurers this was a breach of terms, and Fuller didn't receive a payment because he should have terminated the cover crops before planting corn and soy. At that point, he had no savings and was forced to take out an expensive loan just to keep the farm going. Fuller appealed the decision and won the fight with the insurer, but his life didn't get any easier. In 2014, Fuller had no fewer than seven unannounced farm inspections. Most farmers don't apply herbicides themselves but contract the work out to specialist companies. That year, the inspectors also paid several visits to the contractor who sprayed Fuller's field, time-consuming and bothersome procedures that ended after the company cancelled their contract with Fuller.

Over decades, farmers in the Midwest have grown practically nothing but corn and soy. The vast monocultures have led to soil degradation and increased pesticide pressure, which farmers combat with chemical fertilizers and pesticides. Over the years, ever more frequent applications were needed and input costs steadily rose. Synthetic fertilizer may be able to improve soil fertility, at least for a while, but the agrichemical industry so far has failed to develop anything that can stop soil erosion. At the end of the 1990s, worried farmers started to think about alternatives, and no-till agriculture began to get traction. The annual 'No-till on the Plains' conference became a forum for farmers who wanted to talk shop about no-till, and soon, more broadly, about anything to do with regenerative agriculture, soil health, and cover crops. Gabe Brown, a farmer from Bismarck in North Dakota, was among the early adopters. In the late 90s, his farm was hit several times by catastrophic weather events: for two years in a row hail totally wiped out the entire harvest, while the following spring a blizzard raged for four whole days and killed 14 new born calves. 1997 was a drought year; there was so little rain that the seeds either didn't germinate at all or the seedlings withered soon after – there was no harvest that year. And the following year a sudden summer hail storm devastated the crops. The 'disaster years', as Brown calls them, were horrible, but today he is grateful for the lessons they taught him. They forced him to completely rethink the way he worked. He realized farmers have to work with nature and not against it. His book 'Dirt to Soil'[2] describes how he started to focus on soil fertility and the diversification of his farm business.

---

2   Gabe Brown: Dirt to Soil, Chelsea Green 2018

Today, he produces beef, pork and lamb, vegetables, fruit, nuts, cereals and honey while his son keeps laying hens. All produce is sold directly. In addition, the farm welcomes ecotourists and, in winter, hunters. It is not about yield per acre, but about profit per acre, says Brown.

Gail Fuller still remembers that he first read an article by Gabe Brown in a farming magazine in 2005. 'I read that so often, the pages fell apart. I had to get a new copy,' he says. For him, it was liberating to read about another farmer doing 'crazy stuff', being successful, and even writing articles about it. 'I saw that I needed to look more at the successes than at the failures. Before the article I was on an island, and I was burnt out'. There definitely were other options and other ways to farm, that much was clear.

It was at a 'No-till on the Plains' conference that Brown and Fuller finally met in person. Pretty soon a third farmer joined the discussion: Dave Brand from Ohio, who was one of the first to work with cover crops and under-seeding corn. By now there is a competition going on between the three of them to try new things and find innovative solutions for the existing problems.

It took him several years to realize the importance of carbon, says Fuller: 'Carbon is the driver. It is truly the foundation of life, both above ground and below.' Everything he does is geared towards soil improvement. His guiding principle is to disturb the soil life as little as possible. In an ideal scenario, all agricultural produce would come from perennial plants. With fruit and nut trees, soft fruit and permanent pastures, that is already the case. To him, animals, cattle in particular, are essential in farming. Regenerative farmers like Fuller don't look at yield per acre but understand the whole farm as an ecosystem in which everything is interconnected. Most vegetables and grains are annual plants. Here, too, Fuller is seeking innovative solutions. His 'British White' cattle herd is the first stop on our farm tour. The coat of 'British White' cattle is mostly white, and with their dark or dark-tipped ears and noses the animals have very expressive faces. There are a few brown or completely black cows, too. Every four or five years, Fuller brings a new bull into what otherwise is a closed herd. He believes cattle have to be able to adapt to where they live. 'We bred livestock that didn't know what to graze. They had to relearn to graze mixed cover crops'.

Fuller takes time to observe, to notice subtle changes that may happen over days or even weeks, and that forms the basis of his decision-making. We are standing at the edge of a permanent pasture. It takes Fuller just minutes to move the electric fence and allow the animals to access a different paddock. The old one will be closed off to give the grasses time to recover – it is the type of managed

Farming with benefits

grazing we've seen at Kevin Fulton's farm and the Frasier Ranch. While Fulton and Frasier keep the animals on mixed and short grass prairie, Fuller works with permanent grassland. In spring, the animals are allowed into the 'salad bar': the fields where cover crop mixes have grown over winter, some with ten species or more, including different legumes, radishes, and herbs. Fuller knows each of his animals and how it will react once the new paddock has been opened up. Since we've been standing at the side of the pasture, the lead cow has not let Fuller out of her sight. As soon as he begins moving the fence, the other animals line up behind the lead cow. There is a strict rank order, everyone knows their place, except some young cattle who haven't quite realized what's happening and have to rush to join the back of the queue. As soon as the animals reach the new pasture, they quickly disperse, some jump into the air – a behavior most farmers only see once a year when cattle that have been housed in a barn over winter get their first taste of spring grass. Fuller says his animals have very individual routines: 'They will often feed on one particular plant on their way from one end of the pasture to the other, and on their return, they will feed on another. They know exactly which plants are right for them. It is like a craving we get as humans, when we suddenly know we want one particular food.' To move the herd takes just ten or fifteen minutes. The fence posts have been placed in such a way that it is easy to create ten different paddocks while maintaining access to the water source. Fuller keeps his cattle on grass year-round, and they get no additional feed, which means it takes between 24 and 30 months until they reach slaughter weight.

We get into the pickup again. On the way to the next field, Fuller points out a vineyard owned by his neighbor. Fuller's partner, Lynette, has a full time job in Emporia, but farming is her hobby and what she loves to do in her spare time. She keeps a small herd of sheep with 45 ewes. The sheep can either graze after the cattle or she 'loans' them out to the owner of the vineyard, where they will keep down the grass and the weeds between the vines.

For a long time, Kansas was the second most important wine-producing region in the US, says Fuller. And he grew up in a time when every farm still had cattle, hogs and chickens and the farmers didn't just grow corn and soy, but also sorghum, alfalfa, different grains, vegetables, and fruit, in particular peaches. Today, in this part of Kansas too, agriculture is dominated by soy and corn monocultures, while soils are degraded and have no structure. 'We have a lot of flooding here, even though it rains very little,' says Fuller. In any case, it's the weather. Signs of climate change are everywhere. 'Take this year: in April we had two hard freezes, two weeks after the date when we shouldn't have had frost at all. On July 30th, we got 13 inches (330 ml) of rain, after an eleven-month period that had been classified as exceptional drought'. From July to the end of October, the total rainfall was 26 inches (660ml), with the average annual rainfall for this part of Kansas at 36 inches (900ml). By now, it is not unusual to have half of the annual rain in a period of four to six weeks.

In 1878, the explorer and geologist John Wesley Powell[3] established that the 100th meridian roughly constitutes the border between the rain-rich East of the country and the dry, arid West. This line has since shifted two degrees or 140 miles (225km) to the east, to the 98th meridian; Emporia is on the 96th. Fuller says: 'We are seeing it here. It is scary as hell.'

The consequences of climate change, long periods of drought and rain that doesn't come in moderate showers but as a kind of moving wall of water, have taught farmers like Fuller to make the improvement of soil quality their top priority. A field that hasn't been tilled has a high content of organic matter, the mycorrhizal network is well developed and microorganisms, worms and other soil creatures thrive – such a field will have 'good tilth'. If soils are compacted, the water will run off, whereas if the soil has 'good tilth', the water infiltration and storage rate will be high, with rainwater only slowly percolating into deeper soil layers and finally into the groundwater. The problem most farmers face is that most grains are annuals. To make sure that weeds aren't competing with seedlings for light, water and nutrients, farmers prepare their fields with plow and cultivator. The loss in soil structure is accepted as the price that needs to be paid. Regenerative agriculture has tried to minimize the damage by direct seeding: a disc cuts open the soil, and a blast of air pushes the seed into the opening, which is then closed again mechanically.

Next, Gail Fuller takes us to a field covered in plants with thick, brown stems and a big reddish-brown seed head. It's milo or grain sorghum, which in Kansas is almost as widespread as wheat, corn, and soy. The soil between the stems is covered, and there are only a few patches where only grasses grow and there is no milo. At the end of May, Fuller directed seeded milo into a permanent pasture. For the next seven weeks, there wasn't a drop of rain and he had little hope that any of the seeds would germinate. But then the rain came, and within days, young, green shoots were visible. Not all plants set grains, but many did, and Fuller is pretty pleased. Direct seeding into cover crops isn't easy, but to seed

---

3   https://www.earthmagazine.org/article/dividing-line-past-present-and-future-100th-meridian
    https://e360.yale.edu/digest/a-north-american-climate-boundary-has-shifted-140-miles-east-due-to-global-warming

milo directly into permanent grassland was a real experiment. The established grasses have a deep, dense root network, and any new plant will have a hard time. 'We came very close to having a full grain sorghum harvest,' says Fuller, but in the end, it wasn't worth his while to harvest the milo as cattle feed. Instead, he is letting pigs and sheep get into the field, where they can help themselves to the feed. The sheep have also turned out to be very good at weed eradication: in 2017 there was a lot of ragweed in the field. Then Fuller brought the sheep in, they got fat and the ragweed didn't get to seed. It was gone the following year.

Even though there are a few empty spots in the milo field, for Gail Fuller the experiment has been worthwhile financially, too. The only cost factor was the grain sorghum seed. Despite not using any fertilizer or pesticides he was able to grow a substantial amount of fodder that is now available to the animals in addition to the permanent pasture. It's feed he otherwise would have had to buy in or grow somewhere else. And the milo-grassland pasture has become a habitat for many insects. Fuller kneels down and admires the drops of dew in a spider web sparkling in the morning sunlight. When he carefully pushes the leaves of grasses and herbs to the side, dozens of pill bugs or 'rolly pollies' scurry into the safety of darkness. And there are lots of ants, which is good, says Fuller, as they attack pests and their colonies loosen the soil.

Diversity is Fuller's mantra, above and below ground. The more plant, animal, and insect species, worms and microorganisms that coexist, the better. In a functioning ecosystem, each species creates the conditions that help another to live and flourish. 'I am rewilding the pastures, I am rewilding the cattle, I am rewilding myself,' says Fuller.

A short while later, we are standing amongst hip-high prairie grasses. Three years ago, Fuller seeded warm season grasses into a permanent pasture with cold season grasses. For two years, there was little growth, but in the 3rd year, 'the magic happened', the native plants started re-establishing themselves. In 2018, Fuller went a step further and drilled milo and Turkey Red wheat, the same variety Tim Raile grows in St. Francis. The wheat he directly seeded into the prairie grasses grew well, and he hopes that from next year, harvesting it will be worthwhile. His goal is to get the wheat milled and then sell the flour. Like Gabe Brown, he wants to diversify as much as possible and add value to his produce. In 2000, he was still farming 3,200 acres (1,300 ha). Financial circumstances forced him to give up land and shrink the farm to 400 acres (160 ha). He still argues with Gabe Brown about the ideal farm size: Fuller would like 80 acres more to fix the water cycle. Brown thinks his farm is still too big because he can't farm it alone and needs staff. And Gail Fuller and Gabe Brown have another competition on: the 'chaos garden' versus the 'crazy field'. The question is how much diversity on a given acreage is possible and up to what point it makes financial sense. Fuller used a conventional seeder to seed 20 different crops on a four acre (1.6 ha) field: open pollinated corn, five different squashes, three pumpkin varieties, okra, water melons, shell beans, cow peas, green beans, mung beans, cow peas, and other vegetables. Except for the water melons, which just got too little water during the drought, everything else grew. Three years before, Fuller had planted cover crops on that particular field, some of which had gone to seed. During the drought, the 'volunteer' cover crops made a difference. In particular, hairy vetch grew well. In the morning, the plants caught the dew and that provided enough moisture for the corn to grow. 'The Three Sisters' – growing corn, beans, and squashes together – is an agricultural practice developed by Pueblo Indians in the southwest of the US. Corn grows tall and supports climbing bean varieties. Their roots fix nitrate which becomes available to neighboring plants. Squashes grow between the rows of corn and beans; their huge leaves provide shade and protect the soil from drying out. Probably, there was a 'fourth Sister': sunflowers which provide an ideal habitat for beneficial insects. Next year, they will be growing in the 'crazy field', too.

In October, Fuller machine-harvested the corn, beans, squashes, and other vegetables he and Lynette harvested by hand whenever they needed them for lunch or supper. Up to now, most of what grows on the crazy fields ends up in Fuller's kitchen or that of friends and family. Lynette rolls her eyes when Gail Fuller raves about biodiversity and the joy of deciding in the fields what it'll be for supper. For her, the thing that is crazy in this field is the amount of attention and manual labor that is needed. So, we are talking profit. Fuller tries to direct market everything he produces on the farm and to add value through processing wherever he can. 'Walking the line between supply and demand is tough,' he says. In Yates Center, a small town a 90-minute drive south of Emporia, he has discovered a small 'Mom and Pop' slaughter and processing unit licensed for chicken, cattle, and hogs. Some of the meat is made into bratwurst, which sells well and there is demand for hot dogs, but to produce them would require following a different HAACP protocol and a different workflow, changes that couldn't be made so far. Like Jeter Isley, Fuller is deeply frustrated that there are so many products farms and small family business cannot produce because laws and regulations were written with big companies and industrial kitchens in mind.

One of the barns houses several freezers and large fridges. Customers can order meat, chicken, and sausages by phone or online and collect them at the farm. Shortly after we arrive two women drive into the yard who want to pick up meat for a community BBQ. A few years ago, Fuller got in touch with a CSA program in Lawrence, just outside of Kansas City. Once a month, he delivers orders of meat, chicken, and eggs to the CSA distribution points. Initially, he had hoped that customers might buy a quarter or maybe even a whole side of beef, but most households lack not only the freezer space but also the money to pay for the meat upfront. Fuller now delivers to three CSA schemes serving between 200 to 300 customers, but only between 30 and 40 families order from him regularly.

Emporia has a population of 25,000; Fuller sells meat, produce, and eggs at the weekly farmers' market. But finding new customers is hard work, and costs a lot of time. 'I know how to grow crops and raise animals,' says Fuller. 'I don't know how to market, keep an inventory and advertise. That really is the biggest hurdle.' Up to now, the chickens have proven to be real poster girls for the farm. It is not possible to let them roam free completely as there are too many hungry coyotes, racoons and chicken hawks around. But there are now two mobile hen houses for the 200 laying hens and several 'guard geese' who sound the alarm when there is a predator lurking. The 700 meat chickens are split up in groups of 50, each living in a 8 x 12ft (3x4 m) movable cage. They have an enormously beneficial impact on the soil, says Fuller. It is like mob grazing, but they need to be managed well. Their continuous scratching and pecking can do real damage fast, and they have to be moved once or twice a day. Visitors are fascinated by the mobile hen houses and movable cages, and many of them have become customers. Marketing remains Fuller's biggest problem. Especially in the Midwest, people don't know about good food anymore, he says. Fast food and ready meals which can be shoved into the oven or a microwave, are the norm. 'We need to slow food down'. That's only possible through training. He dreams of transforming the farm into a center of education. He already organizes field days for farmers interested in learning more about regenerative agriculture. Fuller not only demonstrates how he works, what solutions for existing problems he has found and the possibilities that have opened up, but he also invites speakers: soil scientists, dieticians, and other experts to share their knowledge. And in future, he wants to set up agriculture education days for other professions: chefs, lawyers, medical professionals… all of them should learn where the food they eat, prepare or serve comes from, what is available locally, and what is in season and when. He hopes that the participants will communicate and widely spread the message. But his frustration is palpable: 'I am ready to lead the horses to the water, but I can't make them drink.' Organizing education days and workshops are long term projects. What Fuller wants to do most of all is run his farm in a way that plants and animals flourish. For him, both are connected to soil and soil quality: 'We grow soil and out of that grow communities. Agriculture needs to be based on principles, not practices. I want to grow food and soil with integrity.'

Some 40 miles (60km) east of Emporia and Gail Fuller's farm lies Waverly, a small community with a population of just over 500. Darin and Nancy Williams farm 2,000 acres (800ha) just outside of Waverly. Like Gail Fuller and Lynette, they work to produce food that deserves the name. They are committed to regenerative agriculture, and they have a strong sense of community and supporting others. Geographically, Emporia and Waverly may not be far apart, but each farm is different, circumstances vary, and life events lead into different directions and thus to different ways of approaching a problem, different ideas and solutions.

Designer living for laying hens on the Williams' farm

It's the end of October. Behind the Williams' modern bungalow, a large paddock is fenced off. About 50 turkeys run around excitedly, inspecting the visitors. They will roam this pasture for another two weeks before they are slaughtered and dressed in time for Thanksgiving. The life of these birds does not compare in any way to that of the turkeys in industrial units, where thousands of birds are kept in a confined space. The meat quality of free-range turkeys is much better, says Nancy Williams, 'people love the quality, but they are shocked at the price.' On the farm, Nancy Williams is in charge of poultry and sheep. 'At a supermarket, you can get cheap birds for 99c per pound, and ours cost 4,25$

per pound. Many do not understand the difference in quality and the health aspect of good meat.' The mobile hen house is parked in the next paddock. The window flaps of the repurposed trailer are open to allow air circulation. A wide board with tiny steps leads to the entrance and allows the hens to stroll in and out at their leisure. Inside, rows of purpose-built wooden laying boxes have been fitted. Nancy Williams keeps 50 meat chickens and 100 laying hens. As with the quality of free range turkeys, customers don't understand the difference between the runny white and pale yolks from eggs from intensively farmed hens and eggs from Williams' chickens. 'When I bring deviled eggs to the church dinner, they look as if I've used food color, as the yolk is almost orange,' she says. Nancy Williams tries to sell the eggs wherever she gets an opportunity. Four years ago, she got talking to her functional medicine practitioner, and they discussed the connection between health and good quality food. He told her that his patients had little access to good, nutritious food. Nancy Williams knew what he was talking about. When we meet her she is in a hurry. In a rural area like that around Waverly, there are many elderly people who are on their own, living at home, but can't shop for groceries or have difficulties cooking. The county sponsors meals for the elderly on low incomes but relies on volunteers to distribute them. Nancy Williams is one of them. I ask what the food is like. 'Not great, and not very healthy,' she says and pulls a face. But probably still preferable to other options: there are few supermarkets in the area, and most of them can only be reached by car. A lot of rural communities don't have a grocery store anymore, which means that the local gas station is likely to be the only place stocking a limited number of food items – like sweet and salty snacks, pot noodles, and frozen pizza. Her conversation with her medical practitioner about the difficulties to get healthy food in a rural area resulted in a plan: would it be possible to give patients access to fresh meat, chicken, and eggs by selling directly? The medical team loved the idea and the Williams bought a fridge which they set up in the practice. Patients can now pre-order items and pick them up at the clinic on a certain day. Two other functional medical practices have joined the system.

Darin Williams looks after the cattle. He used to run 125 British White, but has reduced the herd to 60. Like all the other farmers we visited, Williams, too, finds it near impossible to sell grassfed beef for a decent price. Marketing is one issue, but he also wants to focus on the beans, meaning he hasn't got enough time to run the cattle as well as he wants to or move them as often as would be necessary in a holistically managed grazing system. For the same reason there are now only eleven sheep left on the farm. Like Kevin Fulton, Williams hopes to find a young farmer who could develop the unused potential of the farm to start his – or her – own business. There is enough permanent pasture and fields with winter cover crops to keep both cattle and sheep. He, too, sees stacking businesses on a farm as the way forward: it creates opportunities for young people who would like to become farmers. And such businesses would be profitable, because the farm would be both diversified and highly specialized. Williams would focus on soybean production, cover crops and soil fertility, while another farm partner could optimize grass management and find a good way to integrate cattle and sheep into a multi-year soy-corn-cover crop rotation. The 2,000 acre (800 ha) farm could provide income for at least two families. But diversification is key because multiple uses create synergies which reduce the work load and save money. Williams gives me an example. He sows cover crops to increase soil quality. In winter and spring, cover crops would also be an excellent feed source for the cattle. The animals would graze them hard, which would allow Williams to direct-seed soybeans. As a result, soil quality would improve, and he would not have to use pesticides to terminate the cover crops in spring. This isn't just better for the environment and human health, but also saves money. The farmer who is running the cattle herd does not have to buy in feed. Manure would be an asset and not the problem, which it is in intensive animal agriculture, where it is collected in lagoons and then periodically sprayed on fields or injected into the soil. Much of the nitrate is washed out into rivers and streams where it causes massive pollution – the dead zone in the Gulf of Mexico is mostly caused by agricultural run-off and leaking of nitrate and phosphate. If cattle are kept on grass or grazing cover crops, however, urine and cow pats are spread across a large area and make the use of chemical fertilizers superfluous. In the scenario Williams describes, both farmers would benefit from what the other does – financially, through a reduction of the work load and the sharing of equipment. Both would have more time to solve problems and perfect both what they do and how they do it. To combine growing soybeans with cover crops and cattle is a relatively simple example, says Williams, but the model could easily be scaled up. Each new element – chickens, sheep, complex rotations, vegetables, trees, solar energy… – could create new possibilities and synergies.

Darin Williams knows how difficult it is to become a farmer. He grew up on a farm in Waverly and helped out during vacations and as soon as he got home from school each day. The farm made little profit, and after Williams graduated from high school, he decided to work in construction. For the next 25 years, he mostly worked on sites in Kansas City, some 80 miles (130km) away. Still, he never really gave up on the dream of becoming a farmer. When he married Nancy in 2006, commuting between Waverly and the different building sites was not an option, and neither wanted to have a long-distance weekend-only marriage. Williams started to think about how he could run a farm more profitably than his father had, and started to gather information. At a 'No-till on the Plains' conference, he met Gabe Brown, Dwayne Beck, a scientist and no-till specialist, and Keith Berns who runs a cover crop seed business. More about him in the next chapter. In 2010, aged 39, he took over the parental farm, sold the plow and sowed cover crops. Within eight years, the soil organic matter content rose from two percent to three or four.

What a farmer decides to grow depends as much on what is calling to him as it does on soil and climatic conditions. Non-GM soybeans are Darin Williams' passion. In the first year on the farm, he decided to conduct an experiment: he split a field and grew GE soybeans in one half and non-GE beans in the other. The GE beans yielded 50 bushels, the non-GE beans 48 bushels, but because of the lower seed and input costs, the non-GE beans turned out to be more profitable, despite the slightly lower yield. And he's never had a problem marketing non-GM beans, either directly or through a broker, says Williams. Most go into export to Europe. In the EU, GE soy and corn are permitted as feed, but more and more consumers want meat, chicken, and dairy products from animals that are raised on non-GE feed. The demand is high and so far cannot be met by growers within the EU. He may soon be producing for an Italian company; at the time of our visit, the final decision was still pending.

For 2019, the propsects for the export of non-GE soybeans had improved further, reported the ag-magazine The Farmer[4] in March of that year. Recently Japan, Korea and Taiwan had passed laws that required all soy products served as part of school meals – soy milk, soy sauce and tofu – to be GM-free. In Vietnam, Malaysia, Thailand, and Indonesia, tempeh, a fermented product that is often used as a meat substitute, has to be produced from non-GE soy milk. The Farmer concluded that these changes in the law alone would increase the demand for non-GE soy by 15 million bushels. In the US, however, more than 90 percent of all soybeans grown are GE varieties. 'We can't get any non-GMO seed,' says Williams. One reason is that the agrichemical companies incentivize the sale of GE seed by giving seed traders favorable conditions if they don't stock any non-GE seed. Because the acreage for non-GE soybeans has shrunk so much, little research is being done, breeders don't maintain and improve existing varieties and scientists don't develop new ones. It's a trend Darin Williams sees as a market

---

4   https://www.farmprogress.com/soybean/year-plant-non-gmo-soybeans

Dicamba does not heed the warning signs

opportunity. He started out with Williams 82, a forty-year-old 'public' variety that yields about 40 bushels per acre. Any farmer can grow a 'public' variety without paying royalties and save the seeds for the following year, should he choose to do so. By now he works with K-State 55/18, a variety developed by Kansas State University that yields 58 bushels. He pays a small royalty to K-State for the right to produce and sell the seed. Other farmers can use the seeds to grow and harvest soybeans. They can save seed for planting on their farm the following year, but they are not allowed to sell seed to others. This kind of contractual regulation is very important: by paying a small royalty, Williams compensates the university for the cost of developing a new variety, but it remains a public good. GE varieties are very different. They are patented and remain the property of the company that developed them. When farmers buy GE seed, they sign a contract that specifies what they are allowed to do or not do – for example, it is illegal to share or pass on GE seeds to others, even for research purposes. The violation of such a sales agreement can lead to prosecution and hefty fines. Agrichemical companies make sure that they are well compensated for genetically modifying seeds: at the time of our visit, according to Williams, a bag of Roundup Ready soybeans was 48$. If the beans have not only been genetically modified to withstand being sprayed with glyphosate, but will also tolerate the herbicides dicamba or 2,4-D, the price increases to 58$. The public non-GE varieties cost between 20$ and 28$.

Darin Williams is a thoughtful man and not easily ruffled – unless you ask him about dicamba. 'I really got on well with all my neighbors until that stuff came on the market,' he says. Dicamba has split the farming community, and it may decide the future of farming in the Midwest. For farmers who want to grow non-dicamba resistant GE varieties and for fruit and vegetable growers, dicamba is an existential threat. The bitter controversy over dicamba began in 2016 and is dividing communities and even families. One confrontation resulted in a farmer being shot dead. Dicamba is a broad-spectrum herbicide that was first licensed in the 1960s. It kills leaf plants and was used outside the growing season to eradicate all weeds in a field before new seed went into the ground. Some farmers decided on a 'fall burn' to kill the weeds before the onset of winter. As with 2,4-D, a herbicide better known as 'Agent Orange' when it was used as defoliant during the Vietnam War, there were issues with dicamba from the start. Like glyphosate, both these herbicides can cause drift damage to neighboring fields. But unlike glyphosate, 2,4-D and dicamba are also volatile; under certain conditions they can become volatile again after they've been sprayed and form a gaseous cloud that can drift with the wind for up to 96 hours. Such 2,4-D and dicamba clouds can cause severe damage to perennial plants like vines or peach trees which will have produced young leaves and new shoots long before corn and soy seeds go into the ground. Glyphosate went on sale in the 1970s. The use of glyphosate went up exponentially when Monsanto developed glyphosate-tolerant GE seeds. Suddenly, farmers could eradicate weeds in summer during the growing season – Roundup Ready soy and corn were engineered so that they could withstand being doused in glyphosate while the weeds turned yellow and died. But it took only a few years until weeds like pigweed, waterhemp and marestail developed glyphosate resistance. Then no matter how many rounds of the herbicide farmers applied, some weeds survived. By now, such 'super weeds' can be found right across the US with some growing so tall and strong that they can only be removed manually by field workers. In 2018, scientists in Missouri found waterhemp that was resistant to six different types of pesticides[5]. That development isn't really surprising: only the strongest plants will survive being repeatedly sprayed with

---

5   https://agfaxweedsolutions.a/2018/06/22/missouri-weeds-waterhemp-confirmed-with-6-way-resistance/

herbicides. They are able to develop seed heads and pass their genes on to the next generation of weeds. No breeder could work more successfully!

Agrichemical companies needed to do something to help farmers out and created GE seeds with multiple herbicide tolerances. Farmers could now mix and apply a herbicide cocktail in an attempt to kill off persistent weeds. By 2016, Monsanto had a new glyphosate and dicamba GE soy variety ready for sale. For many farmers, the Roundup Ready Xtend[6] crop system seemed to come just in time. They planted the dicamba-resistant seeds even though the herbicide that was to go with it hadn't yet been licensed for use. As early as the summer of 2016, the number of reports on drift damage started to mount. Farmers who had sown non-dicamba resistant beans saw plants with curled leaves and discolorations that are typical of dicamba damage. A fierce row over liability and compensation began. With glyphosate, it is relatively easy for a farmer to establish which neighbor probably caused the damage, as the herbicide does not drift very far. But the problem with dicamba is not drift damage but volatility, and with a herbicide that can form a poisonous cloud and travel for up to four days before descending, it is practically impossible to establish reliably who sprayed it, where and when. The insurance damage investigators and extension services were at a loss, and Monsanto denied all responsibility: the farmers had used unauthorized products as the Roundup Ready Xtend herbicide wasn't even for sale yet.

Already in that first summer, it became clear that dicamba damage happened more often in some areas than in others and weather conditions played a major role: humidity and thermal inversion increase volatility. In south eastern Missouri and neighboring Arkansas, such conditions are common, which is why the number of damage reports in those two states was particularly high. In one instance, the dispute over dicamba damage ended fatally. On the 27th October, 2016 the farmer Mike Wallace met with his neighbor, Allan Curtis Jones on a dirt road. Wallace was convinced that Jones had caused the dicamba damage in his soybean field. In the ensuing melee, Jones fired his gun but then called the police to turn himself in. Wallace died before he could be taken to hospital[7]. In December 2016, Jones was found guilty of second-degree murder and sentenced to 24 years in prison[8].

---

6 Roundup is the name for glyphosate and glyphosate tolerant crops, Xtend indicates dicamba/dicamba tolerance
7 https://newrepublic.com/article/152304/murder-monsanto-chemical-herbicide-arkansas
8 https://www.arktimes.com/ArkansasBlog/archives/2017/12/15/farmer-found-guilty-in-shooting-of-another-farmer-in-dicambra-dispute

In 2017, everything was supposed to be different. For the first time, farmers could buy not only Roundup Ready Xtend Crop, but also the specially developed, matching herbicide – according to Monsanto a product with a new formula that markedly reduced the volatility of dicamba. Furthermore, BASF and DuPont developed dicamba herbicides which, according to the companies, were safe to use during the growing season. In 2017, farmers planted dicamba-resistant soybeans on roughly 22 million acres (9 million ha), which is almost a quarter of the 89 million soybean acres planted in the US overall. By the end of June, the damage reports started to come in: atrophied plants, curled leaves and few or no pods. By the middle of October, the EPA had registered 2,708 compensation claims. 1.6 million acres of soybean crops, or about 4 percent of all soybeans planted in the United States, had suffered dicamba damage, the agency said[9]. Kevin Bradley, a plant scientist at the University of Missouri, collected the damage reports and sorted them into geographic regions. The resulting map showed that dicamba had caused damage across the United States, from North Dakota to Louisiana in the south, and from Nebraska in the west to Pennsylvania in the east. Worst hit were the states of Arkansas, Tennessee, and Illinois. Each had more than 400,000 acers (160,000 ha) that showed dicamba damage[10]. Experts estimate that the actual damage is much larger, as roughly only one in ten cases was officially reported[11]. And only damage to soybeans was registered, whereas damage to fruit and vegetables wasn't even recorded. Occasionally, the fate of an individual grower made headlines, such as that of a Missouri peach farmer who had his 1,000 acre (400 ha) orchard irreparably damage by dicamba[12]. Nobody knows how many tomatoes, melons, beans, vines, and squashes were damaged on conventional and organic farms and in orchards as well as in private gardens. The damage to wild plants, shrubs and trees was not recorded either.

In view of such losses, state officials, scientists, and farmers began to debate a way forward. From summer 2017, regulators in several states, including of course Arkansas and Missouri, discussed whether the application of dicamba should only be permissible up to a certain date. The proposal suggested to allow spraying only until the end of May or beginning of June. The agrichemical

---

9 https://www.nytimes.com/2017/11/01/business/soybeans-pesticide.html
10 http://fingfx.thomsonreuters.com/gfx/rngs/MONSANTO-DICAMBA/010051ML3P1/index.html
11 https://www.centerforfoodsafety.org/blog/5257/faqs-about-monsantos-dicamba-resistant-crops-and-xtendimax
12 In 2020, Bill Bader, the Missouri peach farmer won a lawsuit against Bayer-Monsanto and was awarded $265 million by the jury. https://www.reuters.com/article/us-bayer-dicamba-lawsuit/us-peach-grower-awarded-265-million-from-bayer-basf-in-weedkiller-lawsuit-idUSKBN20A0JJ

companies were up in arms and fought back hard: in public meetings, they tried to discredit scientists who argued in favor of a spraying ban, they questioned the number of damage reports filed, and they went to court. Their argument: states do not have the authority to issue regulations, only the EPA can do so. And the companies denied all responsibility, if damages occurred, the applicators were at fault because they obviously had not adhered to label instructions.

In late fall of 2018, the EPA decided to renew the license for dicamba resistant GE soybeans for another two years. In addition, dicamba could only be sprayed by licensed applicators, and companies were to extend and clarify the user manuals. The companies reacted and specified[13] the type and exact position of sprayer nozzles, the height of the sprayer beam and the maximum wind speed and temperature. There is a protocol for how sprayer tanks have to be cleaned after use, and applications are only allowed at particular times of day. Nevertheless, independent experts doubt that more training and closely following all company guidelines will reduce or even totally prevent damage. 'Most weed scientists think the problem is the formulation[14],' said Ples Spradley, pesticide assessment specialist at the University of Arkansas Extension at the Louisiana Technology and Management Conference in February 2019. Their argument: dicamba has an inherent volatility, even under ideal temperature conditions and low wind speed. Even if the application happened in this best case scenario, weather conditions would not remain unchanged over 96 hours. Soybean growers therefore have only one option to prevent their crop getting damaged: they, too, need to grow dicamba-resistant GE soybeans. Even if they don't use dicamba themselves, their fields will be protected from drift. The agrichemical companies should be well pleased – and in light of the army of lawyers who were employed to fight in court for farmers' 'right' to use dicamba, despite warnings by a number of renowned weed scientists and pesticide experts, one could come to the conclusion that it may have been part of the companies' business plan from the start. If so, the strategy is definitely working. In August of 2019, The Midwest Center for Investigative Reporting stated: 'The number of dicamba-resistant soybeans planted in the U.S. increased from about 20 million acres in 2017 to 40 million acres in 2018 to 60 million acres in 2019, according to Bayer. That is about 60 percent of all soybeans planted in the

---

13 https://www.agriculture.com/crops/soybeans/15-factors-for-2019-dicamba-applications
14 https://www.farmprogress.com/weeds/dicamba-largest-target-pesticide-problem?NL=DFP-01&Issue=DFP-01_20190213_DFP-01_968&sfvc4enews=42&cl=article_1&utm_rid=CPG02000002986174&utm_campaign=35962&utm_medium=email&elq2=2566a68d1c7f46a59a52ff284f7de88c

U.S., meaning the rest are not resistant to dicamba and are susceptible to damage if unintentionally sprayed by the weed killer.'[15]

An official account of how many acres of soybeans were damaged by dicamba in 2018 was never published. In June of 2018, Kevin Bradley, plant scientist at the University of Missouri, published an estimate: so far, 383,000 acres (155,000 ha) were showing dicamba injuries. The data was incomplete, wrote Bradley, as not all states had responded to the request for updated information[16].

Another headline gives a glimpse into how bad the dicamba crisis has become. 'Arkansas honey seller faults dicamba in closing'[17], wrote the Arkansas Democrat Gazette on 5th January, 2019. Richard Coy runs the Crooked Creek Bee Co. together with his brother. The family business was founded in the 1960s, and the apiary had grown to 10,000 colonies. Over winter, the Coys kept their bees in Arkansas before transporting them to California for the almond blossom. In an interview[18], Coy said he noted the first changes in 2017; the colonies had been weak, and bee numbers had not increased the way they normally do. When he opened the bee boxes, he found that the bees had stored little honey and almost no pollen. Pollen consists of proteins, nectar delivers carbohydrates, and combined, it's the ideal feed for the bee larvae. In north eastern Arkansas, redvine, a perennial woody vine native to the US, is the main source of pollen and nectar. Coy went in search of the plant and discovered typical signs of dicamba damage: on many plants, the leaves were hanging limply and there were almost no flowers. He estimates that 70 percent of the redvine in the region has died. Up to now, his losses for the apiary amounted to $1.1 million. The Coys did not see how the bees could survive in this part of Arkansas. They therefore decided to move half of the remaining colonies to southern Mississippi, the other half to a place near the Canadian border. In the interview, Coy remained optimistic: at least he could move his bees, whereas vegetable growers, organic farmers and orchard owners suffering losses from dicamba weren't getting a second chance.

At present a number of individual and class action lawsuits are making their way through the courts. Farmers and growers are not only suing the agrichemical

---

15 https://investigatemidwest.org/2019/08/27/despite-federal-state-efforts-dicamba-complaints-continue/
16 https://ipm.missouri.edu/IPCM/2018/6/dicambaInjuryUpdate/
17 https://www.arkansasonline.com/news/2019/jan/05/honey-seller-faults-dicamba-in-closing-/
18 Interview with Melinda Hemmelgarn, Food Sleuth Radio, 4th April, 2019 https://beta.prx.org/stories/271882

companies who produce the dicamba GE soybeans, but also the EPA for licensing the herbicide. At the end of 2020, the license will be up for renewal again.

In the meantime, the agrichemical companies are working to find new ways to make dicamba applications indispensable. In March of 2019, the EPA confirmed that Bayer-Monsanto had applied to have the license for dicamba extended for use on corn as the company was working on a new GE corn variety that would be resistant to XtendiMax, a herbicide that combines glyphosate and dicamba. (The move comes despite the fact that as early as 2018, dicamba-resistant weeds were found in several US states). The agricultural publication 'Progressive Farmer' wrote regarding Monsanto's request to the EPA: 'In documents submitted to USDA in 2015, Monsanto predicted the trait could eventually penetrate 89% of U.S. corn acres, roughly 80 million acres (32 million ha) of corn.'

Bayer-Monsanto aims to have the dicamba-resistant GE corn variety market ready by the beginning or middle of the next decade.

Darin Williams finds the spread of dicamba resistant GE soybeans as it is now, depressing enough. To him, it is also an indication of utter helplessness when farmers think they have no other option but to use a herbicide that will potentially harm their neighbors' crops. For farmers, the margin on GE soy and corn is extremely narrow. Climate change and extreme weather events make field work more difficult and create a lot of problems. Weather-wise, 2018 was a mad year, says Williams, and confirms what Gail Fuller had told us the previous day: there was no real spring, then came a late frost which was immediately followed by a period with temperatures in the high 80s (30C), May was extremely dry, and some farmers had so little soil moisture in their fields that they couldn't plant at all. It started to rain in late June, and the whole month of July was extraordinarily wet. Under such circumstances, many farmers are grateful when agricultural advisors from agrichemical companies offer a packet solution: GE seeds plus herbicide and fertilizer. And agricultural machinery producers offer bigger, faster tractors, cultivators, and combines for farmers to make the most of even the narrowest of weather windows. The results are mostly dire, says Darin Williams. 'Some neighbors disced the corn stubble before planting wheat. Then there was heavy rain and the wheat was just washed away. The bare soil, hardened and now they just have a brick layer on the field.' He shows us the field of another neighbor which looks as if it had been used by some tanks in a military exercise, but not like a freshly harvested soybean field. That farmer deep-plowed, says Williams, then he seeded GE soybeans, and sprayed them with dicamba twice, followed by fungicides and insecticides. In Kansas, the weather in fall is usually dry and sunny, but in 2018 it rained a lot – the neighbor harvested anyway, and his huge combine left deep ruts. The result was a lot of destruction for very little yield – 20 bushels per acre is Williams' guess. And the quality of GE soybeans in general is deteriorating, says Williams, as most have blue marks – purple seed stain, a fungal infection which means a drop in price.

Deep plowing and the use of heavy machinery have destroyed the soil

On the opposite side of the dirt road is one of Williams' fields. In 2018, he decided to use it as a demonstration site for his neighbors. In spring, he seeded glyphosate-tolerant GE soybeans directly into cover crops, and he used no herbicides or other agrichemicals during the rest of the growing season. We walk a few yards into the field, and Williams counts the pods on a few plants – the number of pods per plant give a fairly good estimate of what yield to expect. For this field it will be about 40 bushels. On some pods he finds worm damage, but he doesn't think it will reduce the yield much, maybe a bushel per acre. And the financial loss is negligible, which isn't the case for his neighbor, who invested so much money into pesticides.

'On healthy soil you don't need these inputs as cover crops pretty much control the weeds, so they just don't come up, and the soil retains much more moisture, which helps the young soybean plants to grow,' Williams says. The farm road that borders the field is used fairly frequently, and he is sure that his neighbors have taken a very close look at his beans and that his demonstration serves its purpose: helping farmers to understand what difference good soil quality can make. If a farmer absolutely wants to plant GE soy, that's fine, says Williams, but his field is proof that the use of herbicides is unnecessary. And he gets back to dicamba. He shows me an area at the edge of the field There are large 'bald' patches where either no or small, misshaped soybean plants grow – they have been hit by dicamba that drifted with the wind when his neighbor sprayed his field. In the hot, dry climate of Kansas, volatility is much less of an issue than in Missouri or Arkansas, but that doesn't mean there isn't any damage. The farmers here believe that they absolutely need dicamba because their fields are infested with glyphosate-resistant pigweed and marestail. But even by using every herbicide, fungicide and pesticide available on the market, the best yields are around 50 bushels, with the average in the county just 30 bushels, says Williams.

Darin Williams shows us the root nodules on a soybean plant

We drive to the next field, where he started combining non-GE soybeans the day before. The cut-up straw forms a thick layer on the ground, with the tracks of the combine barely visible. 'Some people call soy-straw trash; I call it protection,' says Williams. In the fall of 2017, he seeded rye which had grown hip-high by May. He didn't mow it, but went straight in with a seeder. The weight of the machine flattened the rye before the soybeans were direct-seeded. The layer of rye grass was thick enough to suppress the weeds, and no agrichemicals were needed. I ask again, just to be sure: no, he needed neither herbicides, pesticides, or fungicides, even though the soybean seed was untreated[19]. It's not about having a 'clean field' free of all weeds. The weeds just shouldn't be so numerous as to harm the yield, because they successfully compete for water, nutrients, and light.

Williams pushes mulched soy residue aside and points to a few tender, green tips pushing through the soil – the rye he seeded in fall is already starting to spout again, and he will not have to sow cover crops. The rye will cover the soil over winter with a green coat, and its thick roots will protect against erosion. The experiences Williams had with soybeans in the summer of 2018 are very similar to those Gail Fuller made with milo. Initially, the soybeans didn't do well, as there was just so little soil moisture. 'It was so dry, I thought there is no reason why we should have a soy bean crop here, but once the rain came on August 1st they took off,' he recalls. And now he has a record yield of 58 bushels, almost double the County average. And the quality is good, while his neighbors all got docked for small beans and bad quality. Because his beans are non-GMO, for which there is high demand and little supply, he gets a premium of 75c to 1,25$ per bushel.

Non-GE soybeans combine a number of factors which are good for farm income and the environment: non-GE seed is considerably cheaper than GE seeds. Yields can be the same or higher, and because of the high demand for non-GE beans, they usually get a premium. Because Williams was able to improve soil quality and direct seed the beans, he could increase yields while decreasing input costs – the plants are less susceptible to pests and diseases, so he needs little or no agrichemicals and fertilizer. His only cost factor is the price of the cover crop seeds, but compared to the benefits, it is a small investment, and if the conditions are right, the cover crops reseed themselves – like the winter rye in one of Williams' soybean fields.

---

19 Seeds are often coated with a layer of pesticides to protect the seedling from being attacked by pests while it is still in the soil. Some seeds like canola are coated with neonicotinoids. It's a systemic insecticide; the plant absorbs it, and it harms or kills the insects that attack its stem or its leaves. Neonicotinoids can also be found in pollen and nectar and are absorbed by pollinating insects like honey bees.

And this really shows the difference between the two approaches: for industrial agriculture, what counts is higher yields. In seed breeding and the development of new genetically modified seeds, the one goal is higher yield, and chemical fertilizer, herbicides, fungicides, and pesticides are the necessary means that allow the farmer to promote growth, solve problems, and achieve the maximum yield – that at least is the promise the agrichemical companies make. The same principles apply for growing corn or wheat as they do for producing milk and meat or the production of any industrial food: you need the right materials, the right machines, and a functioning production line – which in agriculture would be land and buildings for raising animals in confinement.

Regenerative agriculture aims to operate within a functioning ecological system. Farming with nature means maintaining this balance, and the smoother the interaction, the better the odds for the environment and farm profits. In a well-balanced ecosystem, problems cannot be solved with agrichemicals, and that, too, means fewer costs.

Darin Williams is adding value in yet another way: he is focusing more and more on seed production, not just non-GE soybean seed but also specific seed mixes. He is a passionate hunter and has developed a 'deer mix'. We drive to a third field which Williams rents from and farms for the owner. It is a veritable jungle of beets, radishes, buckwheat, sunflower, sun hemp, sorghum sudangrass, pearl millet, and cow peas – an irresistible mix if you happen to be a deer. It's a win-win situation, says Williams. With delicious food on offer the animals are unlikely to do damage to other fields. And the fact that he can almost guarantee that deer will show up opens another income stream: hunters will pay good money to take a shot. This particular seed mix also makes for great cattle feed. Cover crops can be commercially viable, and the soil improvement that goes with it is a beneficial side effect. After a bout of torrential rain in August, most of the ditches around Waverly were filled with muddy brown water because the rain had washed away topsoil. In the ditch along the field with the deer mix the water ran clear – the thick vegetation cover had protected the soil from erosion. Initially, Williams had grown oats for seed production on the field. After the harvest, he seeded the 'deer mix', and next year he will grow soybeans.

Williams sees the future for the farm predominantly in seed production. He has founded a company, Natural Ag Solutions, to better sell his non-GE seeds. For cover crop seeds, he cooperates with a small business in Nebraska. He produces seeds for the company, and he buys seed varieties from them to produce his own mixes – such as the 'deer mix'. How did Keith and Brian Berns, two farmers from Nebraska, get the idea to start a cover crop seed business, and why could cover

Farming with benefits

crops be the salvation of many farms in the Midwest? Answers are coming up in the next chapter.

"Don't farm naked" – plant cover crops[1]

Michael Thompson worked hard to become a farmer

Almena, population 400, is a village in northern Kansas, just a few miles from the border to Nebraska. Michael Thompson has sent me clear instructions how to get to the farm: once we've passed Almena we are to take a left onto an unpaved road and stay on it for a good four miles. The farm house sits right by the road, up on a hill, we are unlikely to miss it. It's rained heavily overnight, and dark clouds are still hanging low, adding drama to an otherwise peaceful agrarian landscape of gently rolling hills. Not that we needed added drama... 'Ice' read the red flashing signal on the dash board. I held on to the interior door handle and contemplated when to take the 'brace position', while Martin demonstrated his skills as dirt road rally driver: go slow, but fast enough to make it up the next hill, keep to the tire tracks if you can and whatever else you do: DO NOT BRAKE! Not that there was any ice, what the car warning system identified as such was the soil layer on the dirt road. During the night, the rain had turned into it mud

---

[1] This slogan was created by Practical Farmers of Iowa, PFI, an organization founded by farmers for farmers, sharing information and undertaking farm-based research in conjunction with Iowa State University. In 2016, PFI used the above slogan to promote the planting of cover crops – we still have the t-shirt to prove it.

that was as treacherous as any black ice. A few hours later, we stand in Michael Thompson's machine shed. It is open to one side and the floor is covered with a layer of light colored, almost white dust. Each step leaves a mark as if we were walking through flour. 'That's the kind of "soil" that gets washed off the fields around here when it rains,' says Thompson. 'Farmers here always joke that the pickup is too clean if you can't write a phone number in the dust on the dashboard.' It's not the kind of remark that makes him laugh; the dirt that goes for soil was the reason that he almost didn't get to be a farmer.

Michael Thompson is tall, lanky, and his shoulders come forward a little, as if he often has to stoop a bit in order not knock his head when passing under door frames. When we finally pull into the yard, he is already waiting for us outside. As in many rural parts of the Midwest, mobile connections are patchy, and he was just about to get into his pickup and drive our way, just in case we had ended up in a ditch and needed pulling out.

As we settle in the living room, Michael's wife, Julia provides tea and coffee – not an easy task, the one-year old twins, Joseph and his (two minutes younger) sister Madison have decided that playing with them definitely should have priority. Their older brother, five-year-old Thomas, is at school. He already knows that he wants to become a farmer, says Thompson and he knows how strong that desire can be. When he was Thomas' age, he, too, knew that he wanted nothing more than run his parents' farm once he grew up. 'I loved being outside, seeing the seasons change, seeing the life cycle,' he says. His parents didn't have much time to look after the children; his mother was a teacher, and his father was busy on the farm. 'I was joined to granddad at the hip, we walked through the fields together, he showed me plants, insects and birds'. His grandfather died when Michael Thompson was seven. As soon as he was old enough, he started helping out on the farm. He was aware that there wasn't a lot of money in farming; the family depended on his mother's salary as a teacher. And it was his mother who sat him down at the kitchen table the day he graduated from high school to tell him that he wouldn't be able to farm; it simply wasn't viable. His dad just sat quietly, remembers Thompson. The year was 1997. He learnt to be a diesel mechanic and then did a four-year teaching degree, all the while continuing to look for options how he might get into farming after all. He knew he wouldn't be able to make any money by farming conventionally; constant tilling had led to soil depletion, and there was no fertility left. Signs of erosion were visible everywhere, especially on the hillsides. When the snow melted in spring, or after heavy rains, the soil was just washed away and turned the water in the ditches muddy. The wind carried the soil off in dusty clouds – without cover, the dirt moved like dunes. Initially, Thompson's description makes no sense, but when we later stand in the fine dust of the machine shed, it becomes easy to imagine that this 'soil' will start blowing even if the wind isn't much more than a gentle breeze.

Thompson started looking for alternatives. While doing his educational degree, he had a 30-minute lunch break; he needed five minutes to eat and spent the remaining 25 reading. And that's when he came across an article on regenerative agriculture. 'It all sounded too good to be true, but I thought if it doesn't work, at least I know what to criticize'.

In 2000, Thompson started farming. His father gave him a few acres to use as a mini field laboratory to try out regenerative practices. The land was on a slope, and the soil was extremely poor, depleted, and with deep gullies caused by erosion. Thompson began by planting cover crops. 'Dad thought it was crazy,' he remembers. 'He told me: you are just spending the moisture.' Traditionally, the Thompsons grew corn, soy, and wheat, but they also kept beef cattle. Instead of using herbicides to kill off the cover crops in spring, he put up electric fences and had the cattle strip the field. His father thought buying fencing material was a waste of money. The poly-twine was flimsy and wouldn't hold; fence posts were needed. Then a cow got out because a deer damaged the fence. His father didn't unhook the electricity and got whacked, says Thompson, and grins: 'Dad learnt why cows respect the fence.' The cattle get to graze only a relatively small area at any given time: that way, he is sure that it will be grazed evenly and the soil is protected. If the animals have access to a large area they will pick and choose and leave what they don't like. And they will stay longer to graze. On degraded soil, their hooves can do a lot of damage fast. Already after the first winter with cover crops in the field, the soil was less hard, and Thompson continued to try things out.

The years from 2003 to 2005 were very dry. The conventionally working neighbors fallowed the land, spraying herbicides six to eight times to kill off the weeds. Thompson planted cover crops and never spent a cent on herbicides, which turned out to be a big saving. The following year, he went to the 'No-Till on the Plains' conference and met Gail Fuller and Gabe Brown. By that time, he had already saved a bit of money to buy some land of his own. Since he started his quest for alternative ways of farming and treating the soil he had read every article about or written by Fuller and Brown, both had become role models with a near hero status. Meeting them at the conference very much changed that perception. 'I just got talking to them,' remembers Thompson. It wasn't about being lectured, everyone was equal, exchanging information and ideas, discussing problems and possible solutions, swapping stories – often about the mistakes they made, often not funny at the time but leading to important insights and lessons learnt for the future. And everyone at the conference shared similar goals: improve the soil, use fewer herbicides and less fertilizer, reduce input costs and increase profits. 'Initially, when I started planting cover crops, everyone thought I was a loon, but then I am not one to sit in the coffee shop in Almena,' Thompson says. Like most of the farmers we met, he knows that there is a lot of gossip about those 'new-fangled ideas' regenerative farmers are putting to the test. For many of them, the 'No-Till' conference quickly became a 'safe space' where, for the first time, they did not feel like total outsiders. Instead, they would meet and be inspired by farmers like Gabe Brown and Gail Fuller, who would tell them all the things that are possible once you commit to regenerative agriculture.

Once back from the conference, Michael Thompson really got creative: sudangrass (sorghum) as a summer cover crop, grazed by cattle, oats and turnips in winter, followed by a summer fallow from June to September and winter wheat. It went really well, remembers Thompson, and his only expenses were the seed costs. Over the summer of 2010 he once again planted warm season cover crops: sudangrass, millets, cow peas, and daikon radishes. The field had a restrictive plow pan[2], and the thick radish roots would push through it and open the soil up. 'My father is a county commissioner,' says Thompson. 'There are always meetings on Monday mornings, and he'll be away from the farm. That's when I try the crazy stuff – he can't undo it.'

In 2013, Thompson organized the first Field Day on the farm, it would be held with Ray Archuleta, a soil scientist and agricultural adviser who has a 'cult' status similar to that of Gabe Brown. Archuleta asked Thompson to dig a 'soil pit', deep enough to expose the underlying rock layer. The pit needs to be wide enough to see the different soil layers and distinguish color, structure and consistency. When Thompson started farming in 2000, the soil had looked the same everywhere. The color was a light beige, similar to the dirt we had seen on the floor of the machine shed. The soil organic matter ranged from 0.7 to 1 percent. The 'soil pit' brought a big surprise: the top soil layer was the color of dark chocolate and the soil organic matter content was now 3.2 percent. The water infiltration rate had gone up from 200-300mm per hour to 900mm. 'We had been no-till for 18 years, and the field had been under cover crops for nine years. It's the living roots that brought the change,' says Thompson. Today, he runs the 5,700 acre farm (2,900 acres arable and 2,800 acres permanent pasture) together with his

---

2 A plow pan is a compacted soil layer. It develops between the loose soil layer that has been plowed and the subsoil layer the plow does not reach.

father and brother – from the rise behind his house, he can see their respective farmhouses, 'the perfect distance for members in a family business', Thompson jokes. And he is the one who takes the strategic decisions as to what the future course of the farm will be. What will be planted depends on the weather and – most importantly – soil moisture: one third of the land will be under cover crops, one third for oats, barley or wheat – occasionally triticale[3] which will be either harvested as grain or used for grazing, and one third sorghum or corn. Thompson calls his approach a 'flexi rotation'. In a typical year, he plants wheat that is ready to be harvested in July, followed by summer cover crops like millets and sudangrass which the first hard frost kills in October and 'turns into standing hay'. In winter, such fields will be strip-grazed; the cattle will get to the feed even through dense snow. The permanent pastures are fenced off, each paddock will have between 200 and 300 days in a year for the grasses to recover. On some of the arable land Thompson will seed rye in fall, which will be grazed from about mid-February to the end of May or until the rye starts to develop seed heads. If there is enough soil moisture, he will then plant sorghum, corn or soybeans. If there is too little soil moisture, warm-season cover crops go into the ground. And in some years, there will be a cover crop just for the soil – some of the fields have a lot of calcium, which binds phosphorus. To have enough phosphorus available in the soil is very important for plant growth and Thompson may plant buckwheat which 'unlocks' the phosphorous.

I ask about yields. For corn, the average for this part of Kansas is 60-70 bushels. Thompson harvests between 100 and 140 bushels. He is convinced that the reason for such excellent yields lies in the quality of the soils and the high carbon content. In the past, there used to be good years on the farm, and bad years when nothing grew. Now there is always something to harvest, and even in very dry years they had a crop while the neighbors had none. The plants are much healthier, and therefore more resilient. The year before, a hailstorm hit in late summer, and his neighbors' crops were flat on the ground, while his remained standing. And there are no longer any soil deficiencies which can result in striped leaves – he hasn't observed any of these typical leaf discolorations for years.

If it was just up to him, Thompson would farm without any chemical inputs, but killing off cover crops at the right time with herbicides is a pre-condition for crop insurance, which guarantees a minimum income from the insured field. Up to now his father and brother have not been prepared to give up this 'safety-net' once and for all, in particular as most banks insist on crop insurance as a loan guarantee. To have access to credit if and when he needs is extremely important,

---

3   Triticale is a wheat rye hybrid.

and that's why Michael Thompson does not give up the crop insurance program altogether. The Thompsons own the land, and there is little debt on the farm, but you never know what the future holds – and to work as part of a team on a family farm means being prepared to compromise. That's also the reason why Michael Thompson agreed to plant some GM corn. His father and brother think it is essential, but he disagrees. The cost of GM corn seed per bag is about $350, whereas non-GM corn seed costs about $150, less than half. And worldwide, the demand for non-GE corn is growing, resulting in higher prices for farmers. But because of the lack of infrastructure, the Thompsons have so far been unable to make use of the non-GM premium. No grain elevator in the area is prepared to store GM and non-GM corn separately and pay a premium. The only option would be to sell the corn directly to a feedlot and deliver it – but financially that wouldn't make sense.

Center pivot irrigation

Michael Thompson has no problem with such compromises. 'Since I focus on the soil I don't have to focus on the agronomy of the crops, I don't need fungicides or worry about macro- and micronutrients – the soil does that for me'. And when he walks the fields, he spots a lot more pollinators and plants he has never seen before. Taking pictures and putting them online for people to help with identification has become a hobby. A creek that used to dry out every summer is now constantly flowing. The water has brought back wildlife, including deer who do a bit of damage but Thompson doesn't mind. He loves to see the good this green oasis does. But what matters most is that five-year old Thomas wants to come along whenever Thompson is out in the field. And he is as fascinated by soil as his father. While other kids baked mud cakes, he dug a 'soil pit' and looked for earthworms. 'He loves earthworms,' says Thompson, and if it is about soil, he wants to come, even if it is a meeting with soil experts in the community hall in the middle of winter.

The fields the Thompsons farm and the permanent pastures are clustered around the farm, but some are more remote than others. At present, the cattle are on permanent grazing land almost ten miles away, and because of the rain we can't go there on a direct route, even in a four-wheel-drive, we would most likely get stuck. The Thompsons run 200 to 250 cow-calf pairs. Usually, they sell calves at 300-400 pounds (140-180kg), but if they have enough forage they will keep the calves in the herd and take them to 800-900 pounds (360-400kg). Thompson would like to split the herd into smaller groups and have them intensively graze small paddocks for a very short time. But because the land is not conjoint, frequently moving the animals would often involve loading them on a trailer and driving them to a different grazing area – that would be a lot of very time-consuming work, and quite stressful for the animals. He therefore keeps the herd together and moves it every one to two days. Moving the fences takes about two and a half hours.

Even as we drive by neighboring fields, it is easy to see signs of soil erosion everywhere. Ahead of us, the school bus fishtails, but as it passes us, the driver gives us a relaxed smile and a wave. 'The school bus is coming from our farm and brought Thomas back,' says Thompson, he has nothing but praise for the driver, a woman, who has gotten kids home safely in a blizzard. We talk about schools and the long distances, ordinary tasks like grocery shopping or a visit to a doctor involve long drives, and in case of a health emergency distances, can fast become a life or death issue. The twins were born in December, and started to develop life-threatening respiratory problems soon after. They had to be transported to hospital by helicopter. They suffered no lasting damage from this very serious and mysterious infection. No one knows what might have caused it. And Thompson tells us about another child in the neighborhood who spent the first three years of her life in a hospital in Omaha. The little girl was born without anus, she has to be fed through a tube, the muscles in her left arm are atrophied, and she misses a thumb. I tell Thompson that his description reminds me of our visit with Marghee Maupin, the community nurse on Kauai. She had told us about children born with similar, severe deformities. Medical professionals like Maupin assume that there is a strong correlation between such birth deformities and the pesticides which, on Hawai'i, are used in such huge volumes, but so far there is no direct scientific proof of such a link. The same is true for the high number of cancer cases, on Hawai'i but also in the Midwest and California's Central Valley – all rural areas with a lot of agricultural production and a high use of herbicides, pesticides, and fungicides. Michael Thompson listens without saying a word. After a pause he tells us that he attended a funeral just a few days ago. The wife of a neighbor died of cancer. She was in her early thirties.

Thompson stops the car on top of a hill from where you can see the road on which we came that morning. His family arrived in Almena from Missouri in 1886. In their old home, close to a river, flooding had been a way of life and that's why they homesteaded on a hill, rather than down by the creek. This part of northern Kansas has mostly sandy loam and it is very dry. His grandfather started farming in 1928, right before the dust storms started. During the depression, he went to Colorado to work as an agricultural laborer in the sugar beet production near Fort Morgan. He returned to his farm whenever he could, and in an attempt to keep at least a small area productive and not fallow all his land. But it was the little money he was able to make in Colorado that fed his family and allowed him to pay taxes so he would not lose the farm. Michel Thompson's grandfather told him a lot about the 'Dust Bowl' years, about poverty and dust storms, about neighbors who had to leave their farms and never returned, about the years in which nothing grew. His grandfather's stories have stayed with him ever since and the memory shaped his thinking about agriculture. His father started farming in 1942, and, like many in his generation, took a very different approach. 'The mentality was: "If you see something green in the fallow period you disk it." The soil was like a fine powder. The thinking was you "fluff it up" and water will infiltrate, and they smoothed out every gully by plowing and disking,' says Thompson. At the time, the farm had a thousand acres (400ha) arable land and 400 to 500 acres (about 200 ha) permanent pasture. They had a 'farrow to finish' swine operation: they kept sows, and the piglets were born and raised on the farm until they were sold for slaughter. By the 1990s, family farms with this type of production system couldn't make any money and were barely able to survive. 'There were two options: invest in buildings for a more intensive hog operation or buy rangeland,' says Thompson. 'We did the latter, we sold the pigs and invested in land.' His father practically worked day and night, but never managed to be profitable, he farmed just to keep the farm. Like most in the region, he blamed drought conditions and the generally low rainfall in the region for all the problems he was facing. Thompson blames bad soil: 'Not that we didn't have enough moisture, but it never infiltrated because the soil was so bad. When it rains, the top turns to pudding, and underneath it stays dry. When it's hot it bakes into a brick.' Conventional agriculture has set a vicious circle in motion, says Thompson. The soil is so depleted that the farmers plow up permanent grassland to grow soybeans and corn. It only takes four to five years until these soils, too, have been depleted, the soil structure has been destroyed, and even large amounts of fertilizer don't deliver a good yield. For many, the only way out is to try and buy more land and grow more, in the hope of making up for the drop in yield. A medium-sized farm used to have 4,000 acres (1,500 ha); now it is 10,000 acres (4,000 ha), and some farm as much as 30,000 acres (12,000 ha).

With cover crops and grazing cattle Michael Thompson has demonstrated that there are other options – even for celebrating Independence Day. 'When I was little, in July everything was brown and we were never allowed to have a July 4th firework because of the fire risk,' says Thompson, 'now we can always have one on the day because there is so much green'. He sees the cattle as a 'tool' to repair the system: 'We need to mimic nature, and grassland and grazing animals have coevolved over millennia.'

By now we are back at the farm. Before we leave I ask Thompson what his hopes are for the future. He would like to have more animals, including pigs. He doesn't like the meat his wife buys at the supermarket, the quality is bad. He would like to farm with fewer chemicals and eventually not use any. But he doesn't aim to have the farm certified organic, he would have less flexibility – 'it would box me in, I would have fewer options'. Another reason not to go for an organic certification is the lack of infrastructure: he would be able to produce organic meat, cereals and corn but he would face the same problems as the organic farmers in St. Francis: without certified organic processors and a marketing infrastructure that guarantees chain of custody, he would be unable to get the organic premium. He hopes that in future regenerative and organic farmers will cooperate much more closely. 'We all are trying to achieve the same goals,' says Thompson, and in his opinion that's a very good basis to create a joint infrastructure for marketing, and to make use of synergies, create brands and get better market access. But all of that is in the future. At present, Thompson tries to convince young farmers in the neighborhood to give regenerative agriculture at least a try. Some had planted cover crops in fall, but the following year they had tilled again – so far, no one has dared to rely on nature. Only one farmer has asked him for advice on cover crops and now grows them very successfully. Thompson is optimistic that others will follow.

Farming with benefits

We've been standing in the yard for a while now, Martin with the car keys in his hand. But new topics keep coming up. To Thompson, regenerative agriculture is so much more than a collection of promising farming techniques. The degraded soil that turns into dust, mud or becomes rock hard, the scars of soil and wind erosion on neighboring fields, these are daily reminders of what's at stake here: the way he farms will decide about his son's future, whether Thomas will be able to realize his dream to become a farmer or not. 'Hold on,' he says suddenly, and runs back into the house. A few minutes later, he comes back and hands me a book: 'Farming the Dust Bowl' by Lawrence Svobida. He has just bought a new copy, he says; he has read the old copy so many times that it's falling apart. But now he wants me to have the book, so that I really will understand what this is all about. And then he gets into his pickup and drives ahead of us through the mud until we have safely reached the main road.

Main street, Bladen, NE

We turn north. Our next destination is a village in southern Nebraska. 'Welcome to Bladen, Est. 1886', a sign tells travelers and visitors. At the beginning of the 20th century, there must have been many of both: Bladen had a population of 900, three banks, a post office, shops, and various businesses. Then came the 'Dust Bowl' decade, and today, a century later, Bladen's population is down to 250. There are still traces of what once was a prosperous small town: some of the old buildings still exist, and if it wasn't for the dusty pickups, it could well be the setting for a Hollywood Western. At the edge of town, a wrought iron gate marks the entrance to the cemetery, founded in 1920. A grain bin – the symbol of intensive, industrial agriculture – forms the backdrop.

Keith Berns – and what soil looks like when you plant cover crops

But there are seeds of hope. Since 2009, one company has created new jobs in Bladen, enough for its location to pop up on Google Maps in an instant. 'Green Cover Seed' – a huge sign indicates where we have to turn onto a dust road. After a few hundred yards, grouped like chess pieces on a board, there are silvery grain bins, shining in the evening sunlight: stumpy, fat silos, small ones on stilts and connected to a conveyor belt, so that their contents can easily be bagged or fill a truck, tall lean bins, towering over the warehouses. There are 37 bins in seven different sizes, the smallest of them holding just 1,000 bushels, and the biggest one 25,000, says Keith Berns, who founded 'Green Cover Seed' together with his brother Brian. As the name indicates, the company specializes in cover crop seeds and cover crop seed mixes. Farmers can choose between available mixes or order a bespoke mix. Around 150 different seed varieties are normally kept in stock, and that's not counting the seeds that you can have coated[4] or uncoated.

---

4   There are a number of reasons to coat seeds. One is their size. With a coating, all seeds have an identical size, and that helps to avoid problems in the seeder. But the coating can also contain fertilizer to boost growth once the seed germinates and pesticides to keep hungry predators at bay. Legumes (like peas, beans, lupines…) are regularly coated with rhizobia, which enable the plants to fixate nitrogen. There are rhizobia in the soil, but having an additional dose in the seed coating gives the plants a head-start.

The Green Cover Seed pest control team

Keith Berns and his family live in a modern farmhouse next-door to the business. In the evening, Berns gets a fire going in the cast iron wood burning stove, and we are sitting in the family living room, talking. The Berns family has farmed near Bladen for over 100 years. The seed business and the farm houses sit on land Berns Senior bought in 1966. All in all, Keith and Brian Berns farm 2,500 acres (about 1,000 ha). Like most farmers in the region the family mostly grew corn, soy, wheat, and sorghum. As 40 percent of the farm is irrigated, yields were consistently good, even in drought years.

Initially, Keith Berns trained as a teacher and an ag educator, and for ten years, he taught at different high schools, until he returned to the farm in 1998. And for him, too, it was the 'No-Till on the Plains' conference that inspired him to rethink how the family farmed and take things into a new direction. Good soil can store much more water than degraded soil, and particularly during the hot summer months, soil moisture is a decisive factor for plant growth. The farm has water rights, which guarantee access to water from the Ogallala Aquifer for irrigation purposes. But between Nebraska and Texas, thousands of farmers are pumping water from this vast underground reservoir. The aquiver stretches over 174,000 square miles (450,000 km2)[5], an almost unimaginable size, but the

---

5   http://www.waterencyclopedia.com/Oc-Po/Ogallala-Aquifer.html

water-carrying rock layers vary in depth, and in some areas the aquiver holds much more water than in others. Most of the region is extremely dry, and there is too little rain and snowmelt to compensate for the water farmers extract. Until irrigation became widespread, farmers in large parts of Colorado, Kansas, and Texas couldn't grow corn and soy. From the mid-1970s, the invention of the 'center pivot' revolutionized agriculture on the High Plains. The sprinkler system, which is up to a half-mile[6] long, moves slowly around a center point and irrigates a circle with a half mile radius. Since then, farmers have been pumping record amounts of water from the aquiver year on year. Anyone who has ever flown from one US coast to the other may have spotted a strange pattern of green circles in an otherwise brown landscape – each of these circles is a crop growing because of irrigation.

At the 'No-Till' conference, Berns, too, met Gabe Brown, who at the time had already worked with cover crops for several years. Among the speakers that year were the soil expert Jay Fuhrer and Adamir Calagari, an expert on cover crops from Brazil. 'I was absolutely fascinated by what they were saying and practicing,' says Berns. Both he and his brother wanted to learn more and find out exactly by how much the water storage capacity of soil improves under cover crops. They decided to conduct a field trial. With the help of some grant money, they were able to buy moisture sensors and planted different cover crop mixes on field strips which were grazed by cattle in spring. They compared the soil moisture with that of strips without cover crops, where only wheat stubble had been left in the field over winter. The result: fields with wheat stubble lost significant amounts of moisture. And: cover crop mixes used moisture more efficiently than single species cover crops.

At the 'No-Till' conference, many farmers had complained that their biggest problem was to find cover crop seeds. After their successful field trial, Keith Berns and his brother set out to find reliable seed suppliers, for their own farm, but also to sell to others. Initially, it wasn't really a business idea but the realization that it makes little sense if each and every regenerative farmer has to do his or her own research to source cover crop seeds and then order relatively small amounts, which increases the shipping costs. When the Berns brothers founded 'Green Cover Seed' in 2009, it wasn't much more than a small space in one of the barns. That year, they sold cover crop mixes for about 1,000 acres (400 ha). In 2010, they bought a mixer and started to do bespoke mixes according to what individual farmers needed and wanted. 'We had a steep learning curve and very

---

6 The half mile length of the movable irrigation system has not been chosen at random, but to fit the average field size, which was initially defined by the 1862 'Homestead Act'.

little time,' says Berns. 'And we had to develop and set up our own system.' Nobody else offered bespoke mixes. 'Green Cover Seed' had to develop their own system for quality control. Seeds have to be tested for germination every nine months, at the latest. Most seeds can be stored for a long time, but if the conditions aren't perfect, they may not sprout once seeded. And that wasn't the only logistical problem that had to be solved.

The next morning, we stand in the huge warehouse where the seeds are mixed and packed. Large containers hold seed varieties that are needed in smaller quantities and wouldn't fill a silo. Next to the office sits a white board that shows which order is just being fulfilled, and a diagram shows the silos and the location numbers that indicate where the required seeds are stored. Among the customers are farmers who need cover crop seeds for a large area but it's also possible to order small amounts, like 20 or 30 pounds for a single acre. Only the handling charges change, it is costlier to mix and pack small amounts than large ones. For a seed mix to be 'stable', it needs to contain at least six varieties. If there are fewer varieties, there will be a danger that the mix 'separates' again and the cover crops are later distributed unevenly in the field. Some mixes contain up to 30 different varieties, but it is of course possible to get seed from just one crop. Varieties differ in price. Rye is fairly cheap, whereas perennial grasses are quite expensive – they are much more difficult to produce as it takes three to five years for prairie grass mixes to get established[7].

To have so many different seed varieties in stock makes 'Green Cover Seed' a high-risk venture. At any given time, the seed inventory is worth between two and three million dollars. And that's the reason why cover crop seeds aren't widely available: conventional seed merchants will sell soy and corn seeds and minimize their risk.

Keith and Brian Berns want to offer a service to farmers across the US, and that is only possible with a large seed inventory. 'Demand varies for different climatic zones, orders come in from February through to November, then things quieten down,' says Berns. By now, 'Green Cover Seed' employs 45 staff. Since 2015, a new software program allows the sales team to see which seeds are available in what quantity. Every order automatically creates a mix sheet with information such as where seeds are located, lot and batch number, origin, last

---

7 Grass mixes can be used to improve permanent pastures. On Mark Frazier's ranch, we had seen an area with cold season grasses, which were an important feed source for the animals. Keith Berns works closely with ranchers and now recommends to seed annual cold season grasses into warm season permanent pastures. In a cooler, rain-rich climate with perennial cold season grasses, he would seed warm season annuals. Berns says it is extremely hard to establish warm season perennials among cool season perennials and vice versa.

date of testing, the germination rate… The label can be printed accordingly. 'The program is saving us a lot of time, we don't have to search anymore, and correct information on what's in stock means that we can place seed orders well in advance and don't run out,' says Berns. Twelve silos can be activated from a single control panel in the warehouse. As if by magic, the right quantity of seed is automatically weighed, placed on a conveyor belt and transported to the warehouse, ready to be fed into the mixer. In all bins, temperatures and humidity are measured continuously . A rise in carbon dioxide indicates a hot spot or the presence of insects. For rodent control Berns, relies on an old-fashioned low-tech solution: a small army of cats patrol warehouses and grounds.

Keith and Brian Berns work with a group of dedicated growers who grow seeds for the company on 10,000 acres (4,000 ha) of land. Most varieties come from US farms, while some clovers are imported from India. Sunhemp comes from South Africa, India or Taiwan, and some vetches can only be grown in sufficient amounts in Australia. On their own farm the Berns family grow rye, triticale, buckwheat, sunflower, mung beans, barley, oats, and vetches. In 2018, 'Green Cover Seed' had a turnover of 15 million dollars and shipped seeds for about 850,000 acres (350,000 ha) of cover crops, while sales were up 30 percent from the previous year.

The 'Green Cover Seed' reputation has long since spread way beyond its home state, Nebraska. I read about the company for the first time in ACRES USA, a US magazine focusing on ecological and sustainable agriculture. In one of the articles, Jonathan Cobb, a farmer from Texas, explained how he was unable to get a cover crop seed mix suitable for the area in which he farms until, after some in-depth research, he found 'Green Cover Seed'[8]. In order to offer a seed mix that is ideally suited for a particular farm, Keith and Brian Berns have developed a web tool. Farmers can access it on the company's website and then call or email to discuss specific questions. 'The most important question is: what do you want to achieve with cover crops?' says Berns. Is it about building soil organic matter? Do you want to suppress weeds? Are you most concerned about pollinators? Does your soil need phosphorous – then you'll need to have buckwheat in the mix; it makes phosphorous available to plants, and it is great for pollinators. Is there a problem with soil compaction? Plants with long tap roots will help to break up compacted areas – radishes, okra, sorghum, sudangrass, and rye are options. Legumes will fix nitrate in the soil. 'If cover crops are mostly grown as fodder

---

8   http://www.ecofarmingdaily.com/interview-forging-better-path-texas-farmer-jonathan-cobb-embraces-shift-conventional-biological-based-practices/

you will want grasses with a brown mid-rib, these varieties have less lignin and for a cow the difference to lignin rich varieties is like for us the choice between an apple and a celery stick,' says Berns. In what climatic region is the farm situated? What is the elevation, the average temperatures, the average rainfall, and what crop is to follow the cover crops[9]? The list of details that need to be considered is seemingly endless, but choosing the best possible cover crop mix will not only determine the yield of the cash crop that follows. At times, getting it right will decide over the survival of the entire business. If cover crops increase the soil organic content by only one percent, 25,000 additional gallons[10] of water per acre can be stored, says Keith Berns.

'We need more research to assess the economic value of cover crops,' adds Brian. More pollinators and beneficial insects, nitrate fixed in the soil, reducing or eliminating the need for synthetic fertilizers, additional feed, weed suppression without herbicides... if more farmers are to work with cover crops in future, they will need to put a financial value on individual benefits.

Berns says he learnt everything he knows about cover crops through farming himself and cooperating with other farmers. Their observations and feedback are essential because they farm under different conditions than he does in southern Nebraska, and to know when mixes do well and when they don't enables him to give much better advice. Berns continues to do field trials on the family farm. He shows us the test plots behind one of the warehouses. Over summer, cover crops grew six feet tall. The first severe frost killed them off, and Berns direct-seeded vetch and rye. The seed drill flattened the frozen stalks into a thick mat. He pulls the layer apart and shows us the first green shoots coming up underneath. Different clover mixes were seeded on the adjacent plots, and there are different alfalfa varieties combined with grass mixes which are either mowed or grazed. On one plot, Berns has combined chicory and plantain; both are deeply rooted and accumulate certain minerals. When they are available in a pasture, cattle will need less salt from a salt lick.

We drive to another field. The land on one side of the road belongs to a neighbor who killed the wheat stubble with herbicides. It rained heavily the day before our visit, and now the ditch running along the field is filled with water. The field on the opposite side belongs to the Berns' farm. In 2017, they planted soy beans

---

9 Soybeans are legumes and fix nitrate. Cover crop mixes preceding them should therefore not contain legumes.
10 In metric measures, an increase of 1 percent of soil organic matter means the soil can hold an additional 2.5 million litre of water per hectare.

followed by triticale[11], which was harvested in 2018. Next, Berns drilled buckwheat for seed production. There is no water standing in the ditch, and the soil is much darker than on the field on the opposite side. Berns pushes his spade into the soil and lifts it up. Suddenly exposed to sunlight, several worms hasten to retreat. And left over from the summer harvest, there are some triticale seeds which have started to sprout and extended tiny roots into the soil.

The longest trial is running on another field. It hasn't been plowed for twelve years, and in that time the soil organic matter went from 1.6 percent to three percent. In 2018, the field was divided into three sections, and Berns has direct-seeded rye and vetch into what remained of the summer cover crops. The three field strips were planted over six weeks in two-week intervals to find the optimal time for drilling. Such details can be extremely important. In a dry region, if cover crops stay in the field too long, they will start to 'consume' soil moisture that cannot be replenished before the next crop goes in. A lack of water at the beginning of the growing season usually cannot be compensated, and will result in lower yields. In areas with a lot of rainfall, however, it can be very beneficial to not graze the winter cover crops and have them grow for longer; they will consume water and thus reduce nitrate leeching.

As for the future: Berns hopes to open 'Green Cover Seed' affiliates further south, in Texas, or maybe Oklahoma. The growing season there is much longer and there is a lot of ranchland. Wherever there are grazing animals, cover crops have a bright future.

Cover crops are part of an agricultural ecosystem. Of course, it is possible to mow them or use a roller-crimper. But in particular with cover crop mixes, up to five passes across a field are needed to make sure that no variety sheds seeds or produces so much lignin that it becomes hard. 'Cattle are still the base for cover crops, they make the system work,' says Berns.

But these days, not every cow remembers that the green stuff on the paddock is edible, and rather lives in hope that the feed wagon will deliver some corn silage instead.

---

11 A cross between wheat and rye.

Farming with benefits

Del Ficke in a field with warm season cover crops

'The genetic side is one of the biggest missing pieces. Cattle are all geared towards feedlot production. In 50 years, we bred all the good out of cattle. It is all about efficiency, how to get the cattle to put on the most weight in the shortest period of time,' says Del Ficke, a rancher, farmer and agriculture consultant. Soil is his passion, and he talks about good soil with the enthusiasm of a chef who has just been handed a case of truffles. A spade, the most important tool for our farm tour, is already in the back of the pickup. The farm is situated in the southeast of Nebraska, just a few miles outside the state capital, Lincoln. In 1860, Johann Ficke left the parental farm in northern Germany and emigrated to America. He first settled in Wisconsin, but he liked the open prairie landscape he saw on a trip to Nebraska, and in 1869, he founded what today is the 'Ficke Cattle Co.'.

Some 70 years ago, the family started breeding Herefords, in addition to growing corn, soy, sorghum, wheat and alfalfa. In Del Ficke's childhood, the family used to farm 7,000 acres (about 2,800 ha) in three counties. Rainfall averages in eastern Nebraska are higher than those in the western half of the state or in western Kansas, but water nevertheless is precious. The Ogallala Aquifer does not reach this far to the east, and wells need to go down 400ft (about 130m). Water efficiency and conservation are very important. 'We have been no-till since

1987,' says Ficke. 'It saves not just water but also time and labor. Our first year without tillage was very dry and we were the only ones who had a crop. Tillage is the most destructive thing you can do to the soil. Chemicals are bad, but they do less damage to the soil structure.'

Del Ficke describes his father as a farmer with an 'academic mind', interested in science and how it might be applied to farming Whenever possible, he sent his son along to farm tours to farms in other parts of the country. 'By the time I was 16, I had seen everything, from blueberry farming in the north to cotton growing in the south'. After graduating from high school, Del Ficke had planned to go to college, but then his father suffered a severe heart attack, and he returned to the farm instead. But his new career as farmer and cattle breeder didn't go to plan either. Ficke developed severe back problems, and had to undergo three surgeries. During the long recovery period, there was a lot he could not do on the farm. He used the time to attend college after all. For the next five years, he ran the administration of a medical clinic, and later worked with an organization that helps farmers with disabilities. Throughout this time, he stayed involved in the day-to-day running of the farm until he was able to return fulltime. Ten years in the health sector, first as a patient and later in administration, gave him time to reflect on many issues. He was able to see farms and farming as it is perceived by non-farmers, and that led him to re-evaluate the role of farming in and for the community. There was no key event or light-bulb moment, except for this one morning when he went for a walk on the farm and things fell into place, he knew what he wanted to do and he knew it was right the right path – for the farm and for himself. 'All of our soil is on life support. Nothing comes as close to an endangered species as soil. Food, family, community, environment - everything depends on soil health,' Del Ficke says. He is convinced that in the long run, farmers have to work with nature, and to him, cattle are an integral part of a functioning agricultural system. And that is particularly true for the Great Plains, which, until roughly two centuries ago, consisted of nothing but seemingly endless prairies and the buffaloes that grazed them.

Today, the farm has just 620 acres of land but a herd of 70 to 100 cow calf pairs and two bulls. Forty percent are permanent pasture, 60 percent are farmed by a farmer who works very closely with Del Ficke and grows corn, soy, and sorghum in rotation. Over winter, he plants cover crops, and every fourth year is 'one for the soil', with summer cover crops that are harvested as fodder. Over several years, Ficke conducted a number of field trials with different cover crop mixes, and by now he knows exactly which mix is ideal for which soil and when.

We get into the pickup and start the farm tour. The first field is in the second year of a five-year trial. Colored markers indicate the line between two field sections, but even without the markets the difference is obvious: on one side green plants stand tall between the dry plant matter left over from the previous year; on the other side only tender shoots are visible, cover crops – vetch, red and white clover, winter peas, radishes, rye, and triticale – are pushing through other plant residue. The summer cover crops, warm season grasses and sorghum, have recently been grazed. Del Ficke has grabbed the spade and is calling us over to the non-grazed side. He lifts a clod of soil, and earthworms hurriedly retreat into their worm channels. 'Doesn't this look like black cottage cheese?' says Del Ficke and beams. The soil sample from the neighboring field looks similar, and the soil is dark with good tilth. The cattle that recently grazed this area have left numerous cow pads; one cow produces 35kg to 45kg dung in a day. Carefully, Del Ficke pushes one of the cow pads aside and lifts up a clod of soil from underneath. There are numerous worms which dangle off the soil like spaghetti. The difference to the previous soil probes is stark but not surprising: the worms have congregated under the cow pad, which, to them, is like a rich buffet. They help themselves and drag their portion through the vertical worm channels deep down into the soil. Their excretions are fantastic fertilizer and a food source for different soil organisms.

We get into the pickup again, and a little later, Del Ficke turns onto another dirt road and stops next to something that looks like a grass forest. Sorghum and other warm season grasses stand six feet tall. Ficke grabs the spade and seeks out a path into the jungle of rustling, brown stems. Depending on their condition, these warm season cover crops can be harvested for seed or used as fodder. In the midst of the dried out, brown plants Ficke discovers radishes and beets – volunteers from last season's winter cover crops. Some of the plants self-seeded and have grown into mature plants with strong, wide leaves. The topsoil layer here is now about four inches thick, says Del Ficke.

A little way up the hill lies a prairie grass meadow with a pond and surrounded by trees and shrubs. Now, at the end of October, the grasses here are dry and brown, too, but it remains a peaceful, idyllic spot. Native prairie grasses have come back. Ficke allows the annuals go to seed, and the result is a mix of warm and cold season grasses. In spring, it serves as 'maternity ward'; the highly pregnant cows and young mothers with their calves are safe from the wind. The cows can choose from a rich mix of sweet grasses and forbs, and they have access to water. What now looks like native prairie as it would have been over millennia used to be one of the worst pieces of land on the farm, says Ficke. The soil organic matter content went from two percent to five percent. And only when the weather is really bad a few additional bales of hay need to be put out.

We walk to another field nearby which is known as 'the salad bar'. His father used to hate this bit of land because even with a lot of chemical fertilizer nothing ever grew there, says Ficke. He stopped using chemical fertilizer 20 years ago, and for the last ten years, the field has been planted with cover crops and grazed. Today, the topsoil layer is 18 inches deep. Almost as an aside Ficke tells us that in 2017 he planted 'blue corn' here, a variety that can have deep blue kernels, milky white kernels with a blue marking, and any combination of the two. I ask him about the back story. 'Blue Corn' is the sacred corn of the Pawnee Nation. The prairies of Nebraska and beyond were their homeland. Over centuries, buffalo meat and corn were staples, but 'Blue Corn' is believed to have medicinal properties, and is considered sacred to this day. Displacement and resettlement forced the Pawnee Nation out of Nebraska, and many settled in Oklahoma, where 'Blue Corn' cannot be grown[12]. Del Ficke is a quiet, unassuming man, but the fact that the Pawnee Nation asked him to grow their sacred corn for them has a deep spiritual meaning for him, too. His connection to a place, to its soil and everything that grows there, his appreciation of what nourishes people and communities, literally and spiritually, is rooted in him as deeply as it is in the Pawnee Nation.

But now it's time to meet the cows! They are grazing on a hilltop, from where we have a great view across the landscape and of the farmhouse. We make heads turn as we walk across the paddock, but after a short inspection, the animals return to grazing or chewing their cud. Del Ficke's cattle are a crossbreed with a colorful family tree, but all animals have a thick reddish-brown coat, some lighter, some darker, and many sport white markings on their faces.

The Herefords the Fickes initially bred are a beef cattle race that originates from England. Del Ficke started cross-breeding in 1984. To him there is little value in breed associations and pure lines. The animals are bred for a certain look that distinguishes them from other races and one criterion only: feed efficiency. 'When it works, it is called line breeding, and if it doesn't work, it is called inbreeding,' says Del Ficke. Because of such narrow breeding goals, the animals started to resemble high-performance athletes who have only one talent: producinge enormous amounts of milk or putting on weight in record time. 'There is a dollar index tied to trait efficiency of gaining weight on a tall frame,' says Ficke, and gives the example of Angus cattle, a breed that originally came from Scotland.

---

[12] https://oklahoman.com/article/5615743/sowing-sacred-seeds-how-the-pawnee-nation-saved-ancestral-corn-by-returning-it-to-its-nebraska-home

'Angus went from being small animals in the 1960s to being giants. They have the capacity to take in huge amounts of feed in a very short time.'. 'Which means they put on a lot of weight, fast.' But 'modern' Angus cannot exist without feed trucks, without people feeding them, without drugs and the care of a vet. 'They really stand on good grassland and don't graze because they are waiting for the feed truck to pull up,' says Del Ficke. Regenerative and organic agriculture needs grazing animals that know how to graze. Cattle coevolved with grasslands; they are ruminants, which means they are able to chew their cud. This process, combined with the workings of their four stomachs and the help of gazillions of gut bacteria, enables them to transform grass into calories. And with that, they become central to a productive and financially viable regenerative or organic farming system – with soil carbon sequestration and climate mitigation thrown

'Cowboy genetics': Del Ficke's GrazeMaster cattle

in as bonus benefits. Planting cover crops, in winter or summer, depending on climate and location, means always having living roots in the soil. And these living roots are keeping the soil ecosystem functioning throughout the year and the soil organisms alive. To protect against soil erosion, conventional agriculture increasingly works with cover crops, too. But in the conventional system, farmers usually use herbicides to 'terminate' cover crops before drilling the next cash crop. On regenerative and organic farms, that job is done by the cows. And Del Ficke breeds cattle who are equipped to do this job perfectly. 'I want the most accountable cow herd, they need to do what I want them to do, and I want the best soil. I will never obtain these goals, but I go out and try,' says Ficke. Over the years, he has brought genetics from a variety of different breeds into the herd, each chosen for particular traits they bring. He started out with Herefords, 'the cow that won the west, they reliably produce a calf every year'. He crossed them with Red Angus, who are very maternal, unlike Black Angus, who can be difficult to handle. Next came Simmentals, popular in southern Germany and Switzerland because they are a 'dual use' breed[13] which can be raised for milk or beef. Del Ficke values a different trait in Simmentals: they use water more efficiently. 'They drink about half of what others consume, which is important in an area where you need to conserve moisture'.. Another breed that contributes important traits is French Aubrac; they forage efficiently and are known for very tender, aromatic meat. By 2014, Ficke had a cattle herd with the right genetics: the animals could stay on grassland year round, and did not need to be housed in winter, they make do with whatever pasture they are being offered, their thick coats kept them warm in winter and the cows are good mothers. The cows have a smaller frame than other breeds, which means they calve with ease and almost never need assistants. And when their meat at some point ends up as a Sunday roast or steak, the quality is such that it's sought after by chefs. 'Cowboy genetics' is what Del Ficke calls the mix. Roughly 20 breeds went into establishing the 'GrazeMaster' brand. Four breeds are used to maintain it. It was a conscious decision not to define the GrazeMaster as a breed. He wants the genetics to remain diverse. The animals should be well suited, happy and striving in different regions, on different forage, and in varying climates. And like Keith Berns from 'Green Cover Seed', who offers bespoke seed mixes, Del Ficke helps farmers to find animals with genetics that make them ideally suited to a particular ranch or farm. And once the choice has been made, 'GrazeMaster' cows and bulls can help to modify an existing herd or build a new one. Ficke usually sells his animals as breeding stock when they are about a year old; the heifers are ready to have their first calf, and the bulls can start serving. Farmers and ranchers can also buy semen.

Del Ficke's second income stream is beef production. Between 15 and 20 animals are slaughtered each year. The meat is marketed directly – in nearby Lincoln and Omaha.

---

13 Cows of a dual use breed will never deliver the extraordinary milk yields dairy farmers have come to expect for example from Holsteins. And they will not put on as much weight as fast as traditional beef cattle like Herefords or Angus. But they bring flexibility. A farmer can decide whether he needs a dairy cow, and a bull calf does not have to be slaughtered after birth because it can be raised for meat.

Farming with benefits

And like all the other regenerative farmers and ranchers we visited, Del Ficke, too, wants to pass on his knowledge and experience and help farmers to realize the vision they have for their business. His job as a consultant begins with a long conversation, usually around the kitchen table, says Ficke. The first and most important question is: what is your dream? Do you want to raise cattle? Do you want to go organic? Is it all about soil? Any business needs to be profitable, but profit cannot be the only goal. 'Farmers suffer from Stockholm syndrome. We've become friends with our captors: agrochemical companies, seed companies, producers of farm machinery, and banks,' he reflects. My hope always is that a farmer agrees to a 10 percent change. I ask them to try things out. If they farm 1,000 acres, I say give me 100 and try something different. Once they see it works, they will want to change more.'

Kick-starting the system

If you are an American reader, forgive me for 'carrying coals to Newcastle' – or is it 'potatoes to Idaho'? Unlike most European readers, you will of course be familiar with the Central Valley and why it matters.

In California's Central Valley water wells reach so deep, oil drill equipment is needed

In the past, the European image of California was mostly shaped by Hollywood and its film industry. Today, tech companies come to mind. Air B&B, Apple, Google, and Uber have transformed San Francisco and the southern peninsula, and most people will have an idea what and where the Silicone Valley is. Fewer, probably, would be able to find Fresno on the map, or Tulare, Bakersfield, Merced, Modesto, or Chico. All these mid-sized cities are situated in the Central Valley, which runs parallel to California's coastal mountain range. The cities may not be well-known, but products from the region can probably be found in each and every US kitchen. The Central Valley is roughly 450 miles (720km) long, and farmers here grow one third of all vegetables and two thirds of all fruit and nuts consumed in the US. According to California Department of Food and Agriculture, CDFA, the value of the state's agricultural produce in 2018 accounted for over

13 percent of the total agricultural value in the US[1]. In 2019, California led the list of the top ten agricultural producing States in terms of cash receipts, followed by Iowa, Texas, Nebraska, Minnesota, Illinois, Kansas, North Carolina, Wisconsin, and Indiana[2].

Central Valley farmers not only supply to supermarkets across the US, but export has become increasingly important, too. Already in the 1960s, in (then West) Germany, raisins from California were considered to be something special. Packed in bright red cardboard, sporting a label with a huge sun and the 'Sun Maid' holding a box of raisins, California's raisin growers had created an easily recognizable brand. In the last two decades, the California Almond Board has been extremely good at marketing almonds from California the world over as the must-have, healthy, super food, as essential as purse or car keys, and to be found in the handbag of any successful, weight- and fitness-conscious woman. Worldwide, 80 percent of almonds are grown in California, and the market for Californian pistachios and walnuts is also growing[3].

Drilling for water

---

1 https://www.cdfa.ca.gov/statistics/
2 https://www.ers.usda.gov/faqs/#Q1
3 Until recently, the product with the highest total agricultural value was milk. For a time, California produced more milk than 'traditional' milk-producing states like Wisconsin and New York. Most of California's milk was exported to China. Overproduction and trade conflicts with China, Canada, and Mexico have led to a severe crisis in US milk production, and many dairy farmers have been forced to give up their farms.

Agriculture in California could be one of the biggest, undiminished success stories ever, if there was enough water. But Californian summers are extremely hot and dry, and nothing grows without irrigation. The winter months are the rainy season, with statistically the wettest months being December, January and February. The snowpack in the mountains constitutes natural water storage. Many hundreds of dams, reservoirs and a complex system of hundreds of miles of canals deliver drinking water to cities like Los Angeles and San Diego and to farms in the southern part of the Central Valley. In the San Joaquin Valley – as this part of the Central Valley is called – farming would be impossible without this water transfer.

One of the many canals that transports water from northern California to the south

There have always been years with less than average rainfall, but drought years have become much more common. The last drought lasted from 2006 to 2016, interrupted by just one wet winter in 2010/11. In particular, in the San Joaquin Valley, farmers relied on drilling ever deeper wells, in fact, so deep that the shallower wells in the surrounding areas fell dry. In a number of communities, people now rely on water tankers for drinking water – and sometimes for any water at all. Water from the aquifer that sits under much of the southern Central Valley has been pumped up in such quantities that the water-containing rock layers collapsed. Subsistence has become a huge issue; foundations of buildings sink, walls crack, the canals leak, roads that once were level dip and rise. Still, the drilling for water continues. Anyone who can't access the aquifer has to rely on – and pay for – water from the canals which 'imports' water from rain-rich northern California to the dry south. Farmers who can neither drill wells nor buy water from the north have to fallow land.

There would be no water in southern California without the Shasta Dam

Not only farmers were hit hard by the drought. We now know that trees, too, suffer irreparable harm during prolonged periods of drought. In 2019, the Smithsonian Magazine summed up the latest research results as follows[4]: since 2011, almost 150 million trees in California have died. The situation is particularly bad in the southern Sierra Nevada. The tree roots may reach 15 to 45 ft into the ground, but in very dry years, even at that depth, there is not enough soil moisture left. The rise in temperatures is the reason why today more trees die than during past droughts. Because average temperatures are higher now, trees need more water overall because they lose more through evaporation from the leaves. Once the soil moisture has been used up, the roots start to die off, the leaves wilt, and the acute water stress makes the trees susceptible to a number of diseases and to the bark beetle. The tree die-off began in the lower and therefore hotter parts of

---

4   https://www.smithsonianmag.com/smart-news/why-californias-drought-killed-almost-150-million-trees-180972591/

the Sierra Nevada, but it has since reached cooler regions at higher elevations, too. Since 2016, up to an altitude of 4,000 ft. (1,200m), about 80 percent of trees have been dying. During the past two winter seasons there has been plenty of rain and snow, but average temperatures have risen. Scientists estimate, that every additional degree Centigrade will result in 15 to 20 percent more dead trees. And dead wood increases the fire risk....

For farmers with fruit and nut orchards, rising temperatures pose an additional problem: the yields go down. Many tree species need a number of 'chill hours', periods with low temperatures, to go into dormancy. Trees hibernate, like a bear would, it is their rest period, and if it is interrupted by too many warm days, the trees will be 'cranky' in spring. Between 1950 and 2009, the number of chill hours dropped by a third, and the consequences are visible in most orchards. In California, cherries, apricots, peaches, pears, apples, pecans, and almonds are affected[5]. Walnuts, too, suffer, says Sean McNamara, who runs an organic walnut orchard together with his father, Craig. 'In addition, we have about 15 percent loss through sunburn. It stresses the trees and makes them susceptible to bacterial blight'.

In the walnut orchard, Sean McNamara works with cover crops and sheep to improve the soil quality further.

---

5   https://baynature.org/article/a-time-of-reckoning-in-the-central-valley/?utm_source=FERN+Newsletter+Service&utm_campaign=1faaddaa74-FERN_Friday_Feed_8_9_2019&utm_medium=email&utm_term=0_c95f7f9b8b-1faaddaa74-120469313

The 'Sierra Orchards' lie on the outskirts of Winters, a pretty town some 25 miles west of Sacramento. We've only been waiting for a minute next to the warehouse when Sean McNamara pulls up in an electric golf cart. It is the most environmentally friendly way to get from one end of the 450 acre (180 ha) orchard[6] to the other. And right now, time is of the essence. It's October, and the harvest is in full swing. The trees are shaken mechanically, and the nuts are swept up by a machine similar to a road sweeper. The yield is not particularly high this year, but the quality of the nuts is very good, says McNamara. Variations in yield are normal; usually there is one abundant year and then a lighter one. The orchard borders on Putha Creek, a river that never dries up completely. Unlike many of the nut farmers in the southern part of the Central Valley where water is extremely scarce, the McNamaras never switched to drip irrigation. 'Sierra Orchards' has many large, mature trees with an extended root system, and they cannot be irrigated by just feeding the roots near the trunk. That's why the irrigation lines are hanging in the trees and sprinklers at the far end of the branches irrigate the circumference of the tree, covering the whole root system. The system looks a bit like strings of Christmas lights with sprinkler heads for light bulbs. This year, it's not just the ongoing harvest that is making fall a busy time, the McNamaras have also decided that parts of the orchards need to be rejuvenated.

Sean McNamara wants to show us what 'rejuvenating' means for an orchard. We get into the golf cart. For a little while, we stay on the narrow, black top road, then McNamara turns onto the grass and into the orchard. The huge, old trees have an elegant magnificence. The sky above us has the intense, deep blue of a fall day, the shadows of the leaves form a delicate pattern on the short grass. It is nice and cool here, and the air carries the earthy scent of wood, leaves, and soil. Except for the gentle purring of the electric engine, it's quiet, we have stopped talking, and are caught up in the beauty of this moment. The magic doesn't last long, as suddenly the aggressive whine of chain saws and the deep din of heavy machinery catapults us back into reality. We reach a clearing where workers are sawing branches off newly felled trees until only the bare trunk remains, while others work with an excavator to get a tree stump and its root system out of the ground. It's an almost primeval scene of carnage and destruction. 'This is an incredibly emotional and stressful time,' says a visibly upset Sean McNamara. The trees in this part of the orchard are 70 years old, which is ancient for a tree in a commercial setting. 'In the southern Central Valley, trees stand for about 35 years before they are replaced, in the northern Central Valley 45 years. Almond

---

6 The orchard has 450 acres of which 375 are in production.

trees have to be taken out after 20 years. Irrigation causes a salt layer to build up, which over the years, damages the roots.'

Craig and Sean McNamara have decided to find a new way to rejuvenate the orchard. It is hard enough to rip out old trees, so it should be done in a considered way that does least damage to the soil and the surrounding ecosystem. Traditionally, a grower decides to get out old trees when it suits his schedule – which is unlikely to be in the harvest season. Huge machines are used to yank out whole trees which are then burnt. Special equipment is used to remove stumps and root systems. The use of so much heavy machinery means the soil gets very compacted. Growers therefore alternate using a deep ripper to loosen the soil followed by disking – the process is repeated until the last remaining bits of wood and root can be raked out. The soil is then leveled – a lengthy process because removing the root systems leaves deep craters, and growers want the ground to be absolutely level and flat. Then, cover crops are seeded, usually a mix of sunflowers and Sudan grass, to dry out the soil as much as possible. Only then are fungicides, herbicides, and pesticides applied. The soil is basically chemically 'deep cleaned'. The purpose of fumigation is to prevent the spread of pathogens, fungi and pests from the old trees to the new ones. Walnuts start to bear nuts from year four and are usually harvested from year seven. But even with this radical method of chemical disinfection, it often only takes a few years until nematodes show up and the young trees have to be pulled out again.

Sean McNamara couldn't bring himself to rejuvenate the orchard with such brutal means. 'In the last three years, we have thrown everything we could think of at the soil to improve the soil quality,' he says. To both him and his father, improving soil quality is the best means against climate change. There is nothing a grower can do to prevent the trees from suffering heat stress and missing chill hours. But soil quality can be improved, and in nutrient-rich soil that can store a lot of moisture young trees have a much better chance to grow and stay healthy. And a healthy tree is better equipped to ward off pests and diseases. 'Sierra Orchards' has been no-till for the past ten years. To minimize soil compaction, heavy machinery is only used where absolutely necessary – for example during the harvest. After pruning, branches and other woody residue is not burnt but chipped and left in the rows as mulch. Over the years, Sean McNamara has tried out different cover crops and has now settled for a 15 species mix that is well suited to the unique conditions of a walnut orchard. And he prepares his own compost tea.

Producing compost in a walnut orchard is very difficult. The hulls surrounding the nuts are particularly rich in tannin, and most growers treat them as waste they have to get rid of. Sean McNamara has developed a method to compost green hulls. Available wood, tree trunks, and pruned branches are chipped and combined with other orchard waste to form the bottom layer of the compost pile. The tannin-rich hulls and damaged nuts are spread on top. This year, sunburnt nuts produced an additional problem because they act like tiny energy cells which overheat the compost. At present, McNamara still works with a 50/50 mix of wood chips and green hulls, but with so many sunburnt nuts, he may have to increase the amount of woody material in the pile. The compost matures for six months, but is only turned during the first two. To increase the decomposition rate, fungi and biochar are added – they are for compost what sour dough is for bread. With a bigger compost turner, the work could be done faster and more efficiently, but they are costly, and right now not in the budget.

The bins have been painted white because that prevents temperatures rising too high inside.

So far, there is no tried and tested method to rejuvenate a walnut orchard and protect the soil. The McNamaras are pioneering their own, organic system, and that's pretty stressful. All trees are cut now, in fall, and before the rainy season begins. At present the area looks like a gigantic, dried out mud wallow for oversized wild boars. The soil has to be tilled and levelled before cover crops can be planted. Sean McNamara has come up with a 'super diverse' cover crop mix, and he wants it to go into the ground before the winter rains begin. In spring it will be grazed by a herd of sheep. Since 2015, each year, a shepherd brings his herd to the orchard in late summer. At the beginning of the harvest, the grass between the rows of trees needs to be short, otherwise the nuts cannot be collected. McNamara would like to have livestock in the orchard all year round. At present, the McNamaras work with two shepherds, and at times as many as 3,000 sheep and their lambs graze between the trees. Health and safety regulations stipulate that the animals have to be out of the orchard at the latest 90 days before the harvest begins. 'We decided that we need more control in regard to times of grazing and stock density,' says McNamara. He's therefore teamed up with a friend who will be the orchard's livestock manager. It's a part-time job which leaves him time to develop his own business. Hopefully, he will be able to build up the 150 sheep into a herd of 750 ewes within five years. The sheep are central to the orchard rejuvenation project: instead of chemically cleaning[7] the soil to prevent the transfer of diseases from the old to the young trees, cover crops will not be grown for just a few months but for five whole years. The cover crops and grass area will be the permanent home of the sheep. Sean McNamara shows us a few trees marked with a blue ribbon. They will remain standing and provide shade and shelter for the sheep in summer. From their 'home' paddock, groups of animals can be easily brought to other parts of the orchard to graze cover crops there. This system provides Sean McNamara with the flexibility he needs, and the livestock manager has enough grazing area for his animals. After five years, the life cycle of pests and pathogens that might threaten the young walnut trees should be interrupted. Nematodes living in the roots of walnut trees have little chance of survival once the walnut trees have gone and there are only unpalatable cover crops to be found. The five-year rest period serves yet another purpose: it will be enough time for soil organisms to break up all remaining wood and root residue and decompose it to a degree that it will no longer be a problem when the new trees go in. Sean McNamara leaves to nature what other growers achieve by deep plowing and pay for with the total destruction of soil structure. McNamara would like to keep not just sheep in the orchard, but cattle and chickens too. Cattle and sheep are complementary in the way they graze, and free-ranging chickens on grassland are the most efficient pest control team anyone could wish for. The 70-year old trees that are being

---

7   Even in organic orchards, the soil can be 'chemically deep cleaned'. Once the young trees have been planted, it will take seven years until the first nuts will be harvested, and if no chemicals are used until then, the orchard can get organic certification.

cut down now will make space for 22 acres (9 hectares) of pasture. But all up, they will have to take out and replant 120 acres. This will be done slowly and for five years, with each section under cover crops and grazed. Keeping livestock in the orchard permanently has become a realistic option. And there are other benefits to using the orchard not exclusively for the production of nuts: it creates jobs and gives another young farmer the opportunity to build a business. In this part of California that's not a small thing: land prices in the Sacramento area are exorbitantly high, and young farmers don't stand a chance against deep-pocketed developers who buy up agricultural land to build houses. And if Sean McNamara gets his way, the orchard will also have space for a resident artist. He wants to cut some of the tree trunks into boards. The old walnut wood could be a wonderful raw material for a 'Sierra Orchards' art project.

Almonds, walnuts, and pistachios are mostly grown in the San Joaquin Valley, the very dry, southern part of the Central Valley. The northern part of the valley is much wetter, and that's why rice can be grown in the Sacramento Valley. After Arkansas, California is the state with the highest rice production in the US. On average, rice is grown on 550,000 acres. Depending on water availability, the acres vary between 450,000 (180,000ha) and 600,000 (almost a quarter million ha). There are about 1,000 rice growers in northern California, only 80 of whom farm organically on roughly 30,000 acres (12,000 ha). About half of that acreage belongs to Lundberg Family Farms and the growers they cooperate with.

From Chico, it's just a half-hour drive to the Lundbergs. The country road runs south, along citrus groves and rice fields. It's early November, and huge flocks of migrating birds are passing through. We see them flying in formation above us and can hear their calls. Towards evening they will settle on the surrounding fields and wetlands to feed and rest overnight. Towards the east, the Sierra Nevada mountains are just visible above the morning mist. Somewhere in the foothills, about ten miles, as the crow flies, lies the small town of Paradise. A fact we only become aware of the next day when clouds of dark smoke shroud the sun and dip everything into an eerie darkness. In the early hours of the 10th November 2018, the 'Camp Fire' erupted. It became the deadliest wildfire in the history of California, to date. By the following evening, nothing but smoking ruins were left of Paradise and two neighboring townships. Tens of thousands of people were on the move, fleeing the flames. It took two weeks until the last fires were finally extinguished and the dead had been counted: 86 people died in the flames, and 18,000 homes and other structures were destroyed. But on this Wednesday morning, on our way to meet Bryce Lundberg, life in California still was as ordinary as it would be on a beautiful fall day.

You can't overlook Lundberg Family Farms. Offices, a shop, the laboratory and the canteen are housed in a spacious, modern and environmentally friendly complex. Warehouses and greenhouses for research and rice plant breeding are situated behind the office building. On the other side of the road, next to the railway tracks, sit several dozen white grain bins in which rice is stored. 'The bins are painted white because we noticed that inside the temperature is 10 to 15 percent below that in unpainted, silver silos,' says Bryce Lundberg. He has made time to speak to us, even though we've arrived in the middle of the rice harvest. The family owned company employs 350 staff. It was founded by Bryce's grandfather, who came to California in 1937. The family had originally settled in Nebraska, but it was at the height of the Dust Bowl era that Lundberg saw no future in farming there and took the decision to move to the Sacramento Valley. The experience of living through dust storms shaped his grandfather and influenced the way he farmed, says Lundberg. When he started growing rice he immediately introduced a crop rotation and vowed never to burn rice straw – a practice that has since been officially banned in California.

The Lundbergs converted to organic as early as 1969 – eating a macrobiotic diet had just become 'a thing', and there was suddenly a demand for brown rice. Until then nobody in California had grown organic rice; there was no established way of doing it and the Lundbergs had to devise their own system, learning as they were going along.

The climatic and geological conditions in California are very different from those in other rice-growing regions. Soils in the northern Central Valley are very heavy clay with a hard pan. Conventional rice growers prepare a seedbed at the end of March or beginning of April by using a chisel plow, followed by four to five passes with a cultivator. Because the soil is heavy it needs to be opened up to dry and loosened to about three feet down, explains Lundberg. The hard pan has to remain intact because it holds the water when the field is flooded. At the beginning of May, the fields are flooded with two to six inches of water. Most farmers seed by air – the ideal period is between May 7th and 15th, as after that, the yields tend to go down. The rice usually is soaked before seeding, the grains absorb the water, become plump and heavy, and will sink to the bottom rather than float on the surface. After that, what's left is to wait for the rice to grow and to spray herbicides as often as needed to keep the weeds at bay.

It's the weeds that make growing rice in an organic system so difficult. In theory, rice should grow on fields that aren't flooded, but that would give rice and weeds about the same chances – which isn't good enough for a cash crop. That's why organic farmers seed rice into water, too. It gives the rice a 48-hour advantage in its race against competing weeds. The water remains in the field for 21 days. In the first 10 to 15 days, the weed grasses grow better than the rice, but after that the rice plants catch up. 21 days after seeding, the rice won't continue to grow while the field is flooded, so the water is slowly drained. The relief will come too late for most of the weeds as they will not have survived the last 48 hours. Water weeds like bulrush, arrow head, and duck salad will have thrived, but they don't stand much chance once the water has been drained. Growing organic rice isn't just a race against time. With the dry phase, the temperature roulette begins. Rice pollen is very sensitive to temperature: it is likely to be sterile if temperatures drop below 50F (10°C), it goes dormant above 95F (35°C) and sterile over 110F (43°C). Rice usually flowers in the first half of August, but in the Sacramento Valley, night temperatures can already drop considerably even though it is likely to be very hot during the day. Both can drastically reduce yields, and the later in August the rice flowers, the higher the risk.

Conventional and organic farming practices differ even after the rice has been harvested at the end of October or in early November. Since rice straw can no longer be burnt, conventional growers flood their fields until the straw has decomposed – an anaerobe process that releases a lot of greenhouse gases. At Lundberg Family Farms, the straw is mowed, chopped and integrated into the soil. Then cover crops are seeded which will start to germinate as soon as the winter rains begin. In spring, they are terminated by mowing and disking. Good cover crops really improve the soil, says Bryce Lundberg, as you are less likely to need a chisel plow. They have tried to mow the cover crops and leave them as a mulch matt. It suppresses the weeds well, but the young rice plants struggle to push through. 'Sometimes it works, sometimes it doesn't,' says Lundberg, and in any case, 'if so much organic matter is flooded, a lot of gases develop.'

With cover crops, crop rotation – every third or fourth year no rice will be grown – and by using chicken manure as fertilizer, the Lundbergs try to further increase soil quality and fertility. And they want to up the on-farm compost production so that less manure will be needed. Over winter, new cover crop mixes are trialled – oats and vetch work well, but the clay soils are too heavy for clovers. Growing beans would be ideal. In cooperation with Washington State University, the Lundbergs are trying to find a Fava Bean variety that does well in this part of the Sacramento Valley. So far, only small bean varieties with a high tannin content are doing well, but they are only suitable as cover crops. However, maybe there are heritage varieties that can be grown as cover crops and for human consumption. Quinoa, too, could be an option. And maybe mustard seed could be integrated into the rotation, says Lundberg. Mustard has anti-pathogenic properties and would 'clean' the soil.

Lundberg Family Farms does a lot of research and runs field trials, often in conjunction with different universities. Improving soil quality is one of the goals, others are climate protection and the search for another cash crop that could be integrated into the rotation or work as a cover crop. Even more important is the maintenance of existing and the development of new rice varieties. The Lundbergs are pretty much on their own with this type of research. Organic rice growing is of little interest to universities and agricultural colleges, and in California, 90 percent of the rice grown is of the 'Calrose' variety. Lundberg Family Farms grow 17 different varieties, among them red, black and striped rice, jasmine, sushi, risotto and basmati rice, long and short grain rice, and even wild rice. Several of these varieties were developed on-farm by a small team of breeders and researchers. Their main work, however, is the maintenance and improvement of existing varieties and the production of enough seed for the Lundberg farm and for organic growers in the region. Organic rice varieties need to have special properties which the conventional ones, which are only bred for maximum yield, don't have. It is near impossible to buy organic rice seed, as seed companies would have to invest a lot of work to develop and produce it: organic rice plants need to show seedling vigor and resilience in order to successfully compete with the weeds that threaten them. Conventional varieties just have to deliver high yields. Resilience is not needed as the farmers are at hand to protect them with pesticides, herbicides, and fungicides.

We are standing in one of the glasshouses amidst individually labeled groups of rice plants. It takes eight to twelve years to select for a true[8], new variety. And existing varieties have to be constantly improved. In organic rice growing, the fields have to be drained very slowly, and that adds 10 to 20 days to the growing season. So, the rice either needs to be planted early or has to have an overall shorter season. The researchers in the Lundberg greenhouses also try to find the ideal water depth in which the rice thrives and the weeds give up as soon as possible. And they are looking for drought-resistant varieties.

---

8 In breeding, plants are selected for certain desired traits. In the best-case scenario, the desired traits are consistently passed on to future generations. In reality some plants will revert back to look more like the original parent plants. A new variety has only truly been established once that doesn't happen anymore and the desired traits are consistently passed on. In Europe, new varieties have to be registered, and for that to happen, the vegetables, grains, or fruit would have to show the same traits. A new carrot variety always needs to look the same, whether it was grown close to the North Sea or near Lake Constance.

For Bryce Lundberg, climate change is the one problem that puts everything else into perspective. He is committed to a carbon farm plan which is already being developed together with Resource Conservation District staff. But what a protocol for climate-friendly, sustainable rice-growing could look like is unclear, because so far no one knows. Lundberg Family Farms already recycle 99 percent of their waste, 25 percent of the energy needs are produced on the farm through solar power, plus there is a sustainability plan with a person to put it in place. Wherever possible, change is implemented fast – once the difference in temperature in white and silver grain bins was established, all bins were painted white. To Bryce Lundberg, research is absolutely central, in the lab as well as in the field: 'We need to learn how different practices affect climate change, what we should do more of and what less. It is imperative that we do this. It is critical for the health of our planet. We need to have a discussion, and we need to make investments even if the accountants look at it differently. Understanding what we do is key.'

Finding new paths –
some ideas and solutions from Europe

It's Monday! Our Riverford veg box is being delivered

I do like Mondays. It's the day our veg box is delivered. Rainbow carrots, true spinach, salad, corn on the cob, pimientos, tomatoes, mushrooms, green beans, zucchinis, onions, and potatoes. It's not just the abundance, but also the colors and the scent of late summer that come with the Riverford Organics veg box – with cabbages, root vegetables and salad, the choice is just as good in winter. 'Live life on the veg' is the Riverford slogan – and on Monday mornings a joyful motto for the week.

In the US, Community Supported Agriculture, or CSA, isn't new. If you live close to a farm that runs a CSA scheme, you can buy a share of the harvest in spring and get a veg box throughout the growing season. Some farmers will have you pick up your box on a particular day at the farm, while others will deliver to a collection point. It's a win-win situation: customers get fresh produce, and farmers receive at least some money at the start of the growing season, which helps with planning what to plant and with budgeting and cash-flow. In a good year, the box will be full and at times overflowing – there probably isn't a CSA customer who hasn't felt a certain despair at the sight of yet more courgettes or squashes in his box. But there are bad years too, when adverse weather impacts growth or a sudden hail storm devastates the harvest. Because they bought a harvest share, customers will have to contend with whatever fruit and vegetables can be salvaged. But even in such a year, the scheme has its advantages: both farmer and customers share the financial loss, and customers get a better understanding of how risky and fraught farming can be.

Grown for taste

Weather risks and marketing are two problems all farmers face. It makes no difference whether they farm in the US, in Europe, Australia, or anywhere else in the world. In the UK, most food is sold through supermarkets, and they have become the gatekeepers who decide which produce gets shelf space and which doesn't. Fruit and vegetables have to be uniform in size and color, and they have to be available throughout the year, at short notice, in large quantities and at a consistently low price. Farmers have to comply, or their produce will not be listed. Usually, only big growers can fulfill such demands, and even they have to deal with contracts being terminated at short notice or produce being rejected. The longstanding debate over food waste has led to some minor changes: some supermarkets now offer 'ugly veg' at considerably lower prices; the produce is perfectly edible, but does not meet all aesthetic standards of size or color. And there now is an ombudsman to whom producers can turn with complaints of unfair treatment by supermarkets. Improvements have been made, but the system at large remains unchanged. Small and medium-sized growers and producers have little chance to get their produce and products onto supermarket shelves.

At the end of the 1980s, Guy Watson was one of those farmers. At the time, he started growing organic vegetables at his parents' farm in Devon in Southwest England. 'Vegetables are my passion,' says Watson. 'They need to taste great, they need to be organic and affordable.' Customers living close to the farm appreciated Watson's produce, but supermarkets showed little interest in organic vegetables. And for regular deliveries to farmers' markets and specialty shops in London, Devon is just too far away. To Watson, a veg box scheme seemed to be the only sensible way to market his produce, and in 1993, he started to deliver to the first 30 customers, most of them friends and friends of friends. Today, Riverford Organics supplies about 55,000 boxes a week to customers in England and Wales. After the coronavirus outbreak in spring of 2020, the number of boxes sold jumped to 75,000.

Watson knew from the start that the CSA model has its flaws: most customers are used to shopping at supermarkets and buying what they want and need throughout the year. A farm can supply produce only during the growing season, and in the UK, the months from March to June are known as 'the hungry gap': root vegetables can be stored, but by early spring the last stocks will have gone, and even vegetable varieties that can be seeded early will be weeks away from harvest. The hungry gap can of course be closed through imports. And different vegetable varieties need different conditions, with some growing better in some parts of Britain than in others. The climate in Devon is mild, and winds from the Atlantic bring plenty of rain and offer some protection from winter frosts, ideal conditions for brassicas, 'But don't try and grow onions here,' says Watson. He figured that he needed to supply customers year round or most of them would drop out of the scheme before the next season. His conclusion: in order to be an attractive option, veg boxes have to be available on a continuous basis and offer a good and varied selection of produce.

As a first step, Watson and 16 other local growers formed a cooperative. Next came distribution and building a supply system – a network of growers in other areas and specialists such as mushroom producers. The Riverford network kept growing: today, there are the Devon growers' cooperative and ten farms in England, six growers in Spain, three or four in Italy, and several in France. A few years ago, Watson himself bought a farm in France to help close the hungry gap. The Vendée, just 250 miles south in western France, enjoys better light and earlier springs, and as a result, the produce from the farm in France can be harvested six weeks earlier than in Devon. 'Our cooperation with farmers and growers is based on long-term relationships and on trust that has grown over many years,' says Watson. Because he is a farmer himself, he knows what is important for a grower: organic agriculture works with long rotations. Farmers and growers need to plan way ahead in the knowledge that there will be a market for their produce. And they need a certain flexibility to be able to react to what's happening in nature – from drought to flood to pest invasion. A growing plan is therefore sketched out as early as 18 months in advance, and everyone, Riverford staff and growers, is involved in the process. Agreements are kept. 'If someone offers to supply us with carrots, but we already have an agreement with a grower, we will stick to it, even if it costs us more,' Watson explains. 'We agree well in advance what produce we will take from which farmer, how much we want and at what price. We may not offer the best prices; if something is in demand, a farmer might be able to sell it elsewhere for more, but at Riverford, we take what we order, even if it can be sourced cheaper elsewhere.' And the suppliers keep their side of the bargain; being able to sell at a higher price one year isn't worth giving up the security of having a long-standing working relationship with Riverford: 'Farmers have peace of mind, because they can focus on growing vegetables, not selling them. They can make longer term investments. It is a great way of doing business and everyone profits,' says Watson. If bigger changes are planned, the growers concerned are informed well in advance. 'We give the producers several years' warning. When we took the decision not to sell anything from heated glasshouses anymore, we told them three years in advance. For some, it was a significant part of their business, and they needed a chance to adjust'.

Taste and aroma are the two main criteria for choosing to grow a particular variety. And with long-term planning, the preferences of the growers can be taken into account. And Riverford encourages farmers to experiment and try out new things – that's how you find out which seed variety does well in which soil or is better adapted to the changing environmental conditions. On occasions, customers find vegetables in their boxes they didn't know existed: cardoons are a staple in northern Italy but are completely unknown in Britain. Kohl rabies, too, need a lengthy explanation as to what to do with them. The Riverford Field Kitchen on Watson's farm in Devon started life as a staff canteen. But the cooks took pride in creating and testing new recipes, and many of them have become part of the ever increasing recipe collection that anyone can access for free through the Riverford website.

Customers can change their orders until 48 hours before delivery. And there is a lot of choice: veg boxes with or without potatoes, without root vegetables, boxes with seasonal British produce only, salad boxes and bags, combined vegetable and fruit boxes, fruit only boxes and bags. Most boxes are available in three sizes, small, medium, and large. Customers can check what's going to be in their box online a week in advance.

In 2014, Riverford added recipe boxes. TV cooking shows are extremely popular. However, studies have shown that many Brits would like to prepare more meals at home, but either lack the time or the necessary cooking skills. A recipe box contains all ingredients needed for three different recipes. Customers should of course have pots and pans, salt, pepper, and cooking oil, but everything else, including spices, mustard and vinegar, will be provided in the recipe box in exactly the right quantities. And not only omnivores are catered for, there are recipe boxes for vegetarians and vegans, too.

Apart from fruit and vegetables, customers can also buy meat, milk and dairy products. There is a small choice of locally produced artisan cheeses, bread from a local bakery, and even olive oil and olives are available. Artisanal products such as cheese are fairly expensive, and it's probably cheaper to buy organic milk and yoghurt from a supermarket. But Riverford does regular price comparisons for produce from three big supermarket chains. The veg boxes including delivery are up to 20 percent cheaper than when buying similar quantities of organic produce in a supermarket.

The boxes are packed in one of two distribution centers, one in Devon, the other in Peterborough, in east England. Riverford employs around 100 drivers, but there are also a number of franchises that organize deliveries in their region and handle customer support.

Fairness is part of the Riverford ethic, with fair pay and fair treatment of partners, growers and the seasonal staff, who often come from Eastern Europe, and have worked on the farm for many years. In June of 2018, Riverford became an employee-owned company, meaning 700 staff own 74 percent of the business, with founder Guy Watson owning the rest. The members' shares are administered by a trust. There are three governing bodies: the elected co-owner council, the board and the trustees.

Riverford delivers to most parts of England and Wales. A third distribution center in Yorkshire had to be closed down for financial reasons. Watson is really sad about that. A logistical setup including an IT system is an investment of two million pounds, and the turnover in the Yorkshire hub didn't merit that. 'I find it quite disappointing. The IT revolution was supposed to lead to greater flexibility, nimbleness, and opportunities for small businesses, but when you look around you, that is not what's happening. Instead, it has accelerated the trend towards ever larger and more concentrated retailers, and that is problematic.'

In Britain, there are now quite a few companies operating online farm shops. Customers can order fruit, vegetables, milk, dairy products and meat from a farm of their choice and the company organizes the distribution. Another business model is the 'virtual supermarket', where customers can buy anything – fresh produce, canned goods, and even cleaning products. To be listed at this type of online shop is about as difficult as getting a supermarket listing, and in Britain, most supermarket chains now have online shops, too, and deliver to the customer's door. In metropolitan areas like London, Amazon Fresh supplies everything, from avocado to tofu and zucchini. 'I am pretty confident in saying that the only ones making any money are the big players like Amazon. I don't think even our supermarkets are making any money out of home deliveries of food,' says Watson. In his opinion, most companies don't treat farmers fairly because they would not be able to be profitable if they did: 'Farm Drop do return a fairly good percentage to their farmers, but they lose an unbelievable amount of money. It's a totally unsustainable business model propped up by venture capitalists who are hoping that companies like Farm Drop will be the next best thing.' Watson believes the trend towards more direct deliveries with ever tighter time slots is a disastrous development. 'The pressure is towards ever greater consumer convenience, 20-minute delivery slots, in some cases the same day you order and evidence would suggest it is a model that's impossible to make money out of. But it's what customers are saying that they want. So, everyone has to do it, that's the kind of mantra in retail, the customer is always right – which is not something I actually agree with – and we say: no, we deliver once a week, your postcode will confirm when a delivery van passes your door, and that's when you get your delivery,' says Watson. 'Ever more frequent deliveries and tighter time slots are the road that is leading us to hell. Environmentally, it is completely unsustainable; companies that do it are probably using three times as much fuel delivering in that system than we do because it is impossible to deliver in any logistically sensible way. The same argument continues through to seasonality of produce. Should people be able to buy strawberries 12 months a year and have them delivered within a 20 minute time slot within 24 hours? That involves a massive environmental and social cost, people working unsociable hours, clogging up our roads with vans going here, there and everywhere. It is totally unsustainable. Consumers don't know the consequences of their desires but I think if they did many of them would be willing to do more compromising.' Riverford has chosen a different path. 'We set our stall out to be different. There is no hope for us to compete with these people, and ethically, I wouldn't want to anyway. Environmentally and socially, it is just madness. We rely on customers who believe in what we do and are willing to make some sacrifice'. For this reason, Riverford is reducing the number of items customers can add to their veg and recipe boxes. The list of over 150 items – from canned tomatoes, jams, chutneys and ready-made foods like dips – has been reduced to about 100. Meat and dairy products continue to be available. 'The focus will remain on fruit and vegetables – that's what we know best,' says Watson, the general message being that we need to eat more vegetables and less meat over all. 'But I still find it hard to imagine a sustainable farming system that does not involve some ruminant livestock. I would be a hypocrite to advocate for a vegan diet.' Watson's brother raises cattle on grass grown as cover crops that are planted in a rotation to enhance soil quality.

Riverford does not sell any fruit or vegetables grown in heated greenhouses and nothing is air-freighted. But in recent years, more produce has been imported to give customers a wider choice throughout the year. At present, the company's carbon footprint is being reassessed. Watson says: 'We might define a certain standard and decide not to sell produce with a higher carbon footprint. On the other hand, every avocado a customer wants and Riverford doesn't sell is a significant drop in order value. We'll have to pragmatically keep pushing at the boundaries'. In 2019, Riverford invested in solar panels on farm buildings, offices, and the delivery hub in Devon. It is the second largest solar energy system[1] in southwest England. All plastic wrappings will be phased out very soon, and by 2023, all deliveries will be done by electric vans.

For years Riverford, has worked hard to reduce food waste. Whether there are more ripe courgettes that even willing customers are ready to tolerate in their boxes or whether the carrots are tasty but stunted in growth because they lacked water in the relevant period of their growing cycle – to turn such vegetables into soups sounds like a great idea. Jeter and Nina Isley on their farm in Kansas had a similar idea and couldn't put it into practice because a farmhouse kitchen is not set up to comply with health and safety standards developed for the food industry. Riverford failed at a different hurdle: customers who buy soups want them to have a certain shelf life and to be available year-round. 'Someone who

---

1   https://wickedleeks.riverford.co.uk/opinion/news-farm-climate-change-renewable-energy/whose-future-their-future

just received four zucchinis in their veg box is not going to order zucchini soup,' says Watson. Soups make sense when there is a glut of vegetables, but that's exactly the time when no one on the farm has time to prep vegetables and there isn't enough freezer space to keep it until needed. For a time, Riverford worked with a commercial soup producer, but the company relied on continuous supply volumes; the production wasn't set up to deal with a glut that didn't fit into the veg boxes any more. This is why processors work with producers who specialize in growing large volumes of vegetable varieties that are particularly suited for the commercial production of soups, ready meals, or for canning. For Riverford, the only solution lies in very careful planning, and of course, staff can take as much surplus or slightly bruised veg as they like, and if there is still some left over, it will be fed to the cows. And maybe, one day, he will found a social enterprise with staff to cut and prep the vegetables and make them into soups and other meals. But first he will have to deal with Brexit. The wrangling that preceded the UK leaving the EU made planning almost impossible. Since the referendum in 2016, the UK agricultural sector has been in turmoil and hard hit, and Riverford is no exception. 'It's been immensely disruptive and has cost a fortune already,' says Watson. The shortage of labor is an additional problem. Ninety percent of seasonal workers came from Eastern European countries, with Riverford working with the same teams for many years. In 2019, only half of the workers returned and it is unclear what will happen once the transition period ends. No one knows how many seasonal worker visas will be issued, and the number the government proposes wouldn't cover even a fraction of the demand.

But these are home-made British problems. The Riverford model gives farmers a say, market access and good prices, while supplying consumers with affordable, high-quality, tasty, fresh, and seasonal organic fruit and veg. It's a concept that could and should be copied in other European countries, and it could work in some more urban parts of the US, too.

This type of direct farm-to-home delivery needs a good transport infrastructure. And the shorter the distances, the better. Many regions in western Europe tend to be overpopulated, and the urban spread makes selling farmland a tempting prospect: for many farmers, getting out of agriculture and selling land to a developer will make a lot of financial sense. Anyone who continues to farm will have to increase productivity and profitability. Both are only possible if soil quality can be maintained or regained – from the farmers' perspective, the fact that soil health is also an active contribution to climate mitigation is just a beneficial side effect.

That's why agroforestry systems have become a hot topic. Silvopastures are nothing new. On the contrary, sheep or cattle grazing under and between orchard trees is a very traditional form of use. In modern orchards with densely planted rows of dwarf trees, there is no space for animals, although the practice continues in old orchards with trees on tall root stocks. But with new technologies and precision farming, silvoarable systems – the combination of trees and cereal production – have become possible. Several British universities have been trialling agroforestry systems for some years. In 2009, Stephen Briggs set up a silvoarable system on his farm near Peterborough in east England. It is the biggest and longest running commercial system in the UK.

Stephen Briggs works with fruit trees in an arable agroforestry system

We're at the end of August. Stephen Briggs scans the horizon where dark clouds are looming. He finished the grain harvest the day before but the straw still needs to be baled and he's not keen on rain just yet. But it does look as if a thunderstorm is heading our way; the wind has picked up and suddenly drives a cloud of dust towards us. It's the soil from a neighboring field that has just been plowed, explains Briggs. 'Look at that,' he says. 'That's why I have planted all those trees.'

Whitehall Farm lies just south of Peterborough; the land here is flat and the soil fertile. In 2007, Stephen Briggs and his wife Lynn were chosen from a field of

85 applicants to be the new tenants of this 102 ha Cambridgeshire County Council farm[2]. Normally, a tenancy is up for renewal every three years. Not a good option for the Briggs, who wanted to switch to an organic farm system, but the landlord got behind the idea and granted them a 15-year tenancy, which has just been renewed for another 15 years.

Neither Stephen nor Lynn Briggs come from farming families. Stephen's background is in engineering. When he found out he didn't particularly like his job in the car industry, he decided to retrain. He studied agriculture and did a master's in soil science. Lynn, too, is a soil scientist. Eventually, Briggs used his Nuffield scholarship[3] to research agroforestry. He visited agroforestry projects in the US, China, and seven European countries. Today, Briggs not only farms, he is also an agricultural consultant, an adviser to the committee for peat bog conservation in Britain, and vice chair of the European Agroforestry Federation (EURAF). But to him, agroforestry remains a means to an end, his priority is soil quality and carbon.

'Everyone said it would be impossible to manage this farm organically and make a profit,' says Briggs. Previous tenants had grown wheat, oilseed rape, sugar beets and potatoes continuously for 40 years. 'The soils were very, very tired, which is a polite way of saying they were knackered'. Strong winds are common in the area and preventing soil erosion was therefore top on Briggs' list of priorities. Rows of trees are ideal windbreakers, but planting them in the middle of a cornfield potentially could cause quite a few problems. That may be one of the reasons why silvopastures – trees planted on grazing land – are much more common than silvoarable systems – trees on arable land. Sheep or cattle can easily move around trees, with a tractor or a combine it's a lot harder.

After Stephen Briggs got permission for the tree project from the landlord, he had to design the system. He decided to plant apple trees on semi-dwarf rootstock because such trees don't grow too high and provide a return in a relatively short time – which is important for a tenant farm. And apples are good for adding value, for example, by producing apple juice.

---

2 Why does a County Council own farms and farmland? At the end of the Second World War, many of the returning men couldn't find work. Britain also faced food shortages; rationing only ended in 1954. Since then, councils have sold most of the land. There are still 250 council-owned farms in Cambridgeshire, the largest number in any English county.
3 A renown scholarship that allows for overseas travel to do research 'with a view to developing farming and agricultural industries'. http://www.nuffieldscholar.org/

In 2009, Briggs converted 56 ha, or half the farm, into a silvoarable system by planting 4,500 trees in 45 northeast/southwest facing rows. The strip under the trees is exactly 3m wide. The alleys between the rows of trees where the arable crops can be grown measure precisely 24m. All of Briggs' machinery is 6m wide and operated on tramlines in a controlled traffic system: because all the machinery has the same width, the same wheel tracks can be used. Once established, the tramlines – or permanent wheel tracks – protect the trees and prevent the compaction of other parts of the field.

Briggs chose 13 different apple varieties, six heritage varieties, six modern ones and Bramleys for cooking. The trees were planted into a geotextile layer to suppress weeds. 'It seemed like a good idea at the time. In hindsight, I wish we hadn't done it. We are learning the hard way,' says Briggs. The geotextile is providing ideal living conditions for mice and voles. But there is an upside: the five nesting boxes for barn owls are all occupied, probably thanks to the smorgasbord the owls find along the trees when the rodents come out at night.

All trees were planted next to a 60mm post and surrounded by a wire guard. The initial cost for the 56 ha silvoarable project was £65,000 – which explains why Briggs didn't plant trees on the whole farm: it would have been too expensive.

By now, the trees are mature enough to produce apples. Seasonal workers are needed for the hand-harvest. Throughout the year, some maintenance work has to be done: pruning, mowing the grass under the trees and training the roots to grow down by cutting to a depth of about 25cm along the edge of the grass verge. The tree roots need to grow deeper, otherwise they might hinder the root development of the annual grains, which will be planted in the alley between the rows. And the grains benefit visibly from the tree mycorrhiza that extend into the crop area. Briggs says the grains growing closer to the trees develop a little bit better and faster than those in the middle of the alley where the mycorrhiza don't reach.

As positive as it all sounds – an initial investment of £65,000 is a lot. Are the trees worth it? They definitely are, says Briggs, and because trees grow only on half of the farm he can make a direct comparison. The arable yields per hectare are the same, and the trees only take up eight percent of the available space. If he can press the apples into juice, the trees are as profitable as wheat or oats would be on the same acreage. And growing two crops instead of one is a good insurance policy – in a year in which wheat or oats aren't doing well or prices are low, he still stands a chance to make a decent profit from the apples. And vice versa. 'I pay rent for every acre of soil,' says Briggs. 'The space above is free, so I might as well use it. Trees take farming into the third dimension.'

I ask about crop rotation. 'It's a "flexi-rotation",' says Briggs, who likes to leave the decision about what to plant where as late as possible. There are a number of important factors: how much did the last crop yield? Is there weed pressure[4]? What are the results of the soil analysis? And how are markets developing? Briggs says he is an 'opportunistic vegetable grower' as it's worth his while only with a good advance contract. In the past, he has grown leeks, beetroot, onions, and broccoli. In 2013, broccoli proved to be very profitable, whereas the following year he made a loss. It was cold in spring, then it got very hot, and suddenly nobody wanted broccoli. Time and again, market conditions come up in our conversation. For several years, Briggs very successfully grew seed grain. Then the customers demanded that the grain was to be stored on the farm until it was needed. Briggs would have had to invest in new storage facilities, plus storing seed grains carries a big risk – from possible water damage to pest attacks. He switched to organic wheat and teamed up with a local windmill. The millers wanted to mill Cambridgeshire-grown wheat into flour for bread that would be marketed as 'grown, milled and baked in Cambridgeshire'. But in a windmill, space is limited. The wheat had to be delivered in 25kg bags, and the cooperation failed because of a lack of infrastructure: there was no processor to be found to clean and pack the grain.

At present, he grows certified gluten-free oats. Customers with gluten allergies rely on certified products. Gluten-free oats fetch a good price, but to achieve certification, Briggs can no longer grow wheat because it could lead to contamination[5]. Briggs uses every available business opportunity: he sells organic straw to a local organic farmer who needs it for growing mushrooms. Once it's been used, he buys it back as fertilizer. By reimporting the straw, he prevents the farm from losing biomass and carbon. 'Everything is about carbon,' says Briggs, which is a relatively recent insight even for a soil scientist. None of the standard textbooks on farm management he used at university even mentioned carbon. Now he wants to get carbon into the discussion and onto the balance

---

4   A lot of weeds in a field are an indication that they've found good living conditions. Organic and regenerative farmers try to choose a crop that will change these conditions. Established weeds will do less well, and before different weeds can establish themselves, yet another crop will be planted, changing the habitat once again. In an ideal world, the crops in the rotation stay one step ahead of the weeds. That's the theory at least. In reality things don't always go to plan. Yet this type of weed control is far superior to the use of herbicides, which will result in herbicide resistance, sooner rather than later.

5   Before and during the harvest, grains can be shed. Some will germinate the following year and grow into plants which might then be harvested together with the oats. For someone with a severe gluten allergy, even such trace contamination could cause a problem. The certification guarantees that the oats are grown under conditions that prevent contamination.

sheet. On his farm, he has to deal with two very different types of soil. One part is peat with 23 percent soil organic matter. There are plenty of mycorrhizal fungi, but the soil lacks bacterial life, and weed control is difficult. The rest of the farm has heavy brick clay. It's easy to spot where the peat ends and the clay starts: the trees have all been planted at the same time, but those growing on clay are much shorter and not as strong.

When he took on the farm, the surface was nothing but a layer of dust, and underneath the soil was so compacted that even with a lot of force it was impossible to get a spade in deeper than a couple of inches. For two years, Briggs planted cover crops to improve the soil, which now has good tilth – the content of soil organic matter has increased, and Briggs pushes the spade down with ease.

In 2019, most of the farm income came from the oats, as the apple harvest was dismal. A number of factors are to blame, says Briggs: the beekeeper who stationed his hives on the farm fell ill, and there were noticeably fewer bees in spring. Below-average rainfall caused drought stress, which was made worse by some nutritional issues, and Briggs says he had too little time to do a good job at pruning. Which isn't surprising: in 2017 Stephen and Lynn Briggs opened a farm shop and have since added a café, a huge project that cost not just money but lots of time and energy, too. The benefits for the community, however, are considerable: Briggs and his wife have created nine part-time jobs. For customers, the visit to the farm shop is a real outing: they can buy fresh fruit, vegetables, meat, dairy products, bread, and other baked goods, and then have lunch or tea in the café where salads, sandwiches, burgers and cakes are on offer. The prices are very reasonable – a typical Sunday roast with all the trimmings costs just under £10 per person. The café has become a destination and meeting point: as we wait for Briggs, some regulars arrive – a group of young adults with downs-syndrome and their carers often come for coffee and cake. Next door, the new education center is almost finished – the rooms can be used for conferences, courses, meetings, events, or lectures. Cambridgeshire County Council paid for the outer structure, but Briggs and his wife had to shoulder the costs for all fittings, from wiring and plumbing to fitting bathrooms, the kitchen, installing coolers, shelf space, and furnishing the café.

The shop sells conventional and organic produce. 'If we had a sign out saying "organic", people would just drive by,' says Briggs. Organic food is still considered elitist, too expensive, and over-hyped by urban liberals. Since the opening, both shop and café have gained a solid customer base among local shoppers. Organic and conventional produce are clearly marked but sit side by side; visitors can see that organic fruit and vegetables are not necessarily more expensive. Stephen and Lynn Briggs have put a lot of thought into the setup and design. Their purpose is to inform and offer a range of fresh produce and healthy, affordable meal options in a friendly, welcoming environment. It's about providing a service to the local community. At the back of the café is an outdoor seating area and a petting zoo with geese, ducks, sheep, chickens, and guinea fowl.

But a farm is an enterprise and needs to be profitable. Today, Whitehall Farm is a diverse operation, but land and farm buildings have a huge potential, more than Briggs can utilize. A cooperation with other farmers or artisanal producers would make sense. Briggs is a very experienced and successful grain grower, but looking after fruit trees needs a completely different skill set. If he could, Briggs would leave apple production and tree care to an orchardist. It could be a great opportunity for a young farmer to get started. Stacking businesses on a farm makes sense, but the tenancy agreement does not allow him to sublease.

But even with a less than great apple harvest, the trees have certainly contributed to the profitability of the farm in 2019: in early August the area around Peterborough was hit by a two-day storm. The grain was already very ripe and the wind literally thrashed it. Briggs estimates that he lost about 20 percent of the yield on the unprotected part of the farm. On the silvoarable side he lost only about then percent. And for organic oats that's a difference of £150 per hectare.

To establish an agroforest system on rented land means being willing to compromise. Planting fruit trees on dwarf rootstock makes sense because they will only take three to four years until they bear fruit. In an ideal scenario, an agroforestry system would combine different tree species, such as fruit and nut varieties and slow growing trees that eventually could be used for timber. And sometimes, it's possible to secure the long-term use in the planning phase. Briggs tells me about a farmer he advised who planted Sichuan pepper, elderflower, and apples because he was able to secure long-term contracts with three different companies.

Technology can make a real difference. Stephen Briggs regularly uses a number of apps and recommends them in his consultancy work. Diversified farms with long rotations, agroforestry systems, livestock, stacking enterprises, and working with farming partners are likely to be sustainable and financially viable. But such operations are also incredibly complex. And they wouldn't work without technology. A silvopastural system such as Stephen Briggs's would be impossible without GPS guided tractors and tram lines. Apps can deliver very precise, localized weather forecasts, others enable a farmer to identify pests or the first signs of a plant disease on his smartphone in the field or to enter data into a soil map as he is out and about taking soil samples. 'Canopeo' is based on research done at Oklahoma State University. It allows the exact measurement of plant growth and better management decisions: once a farmer knows the grass growth data for a particular pasture, he can calculate the optimal number of animals that should graze there and for how long. The Farm Crap App calculates how much carbon goes into the soil through manure and how much is exported through crops. Marketing apps show trends and help with planning decisions because they deliver a better overview. Universities and technology firms develop digital tools; the first robots are being field-tested, artificial intelligence and agricultural machinery fitted with cameras open up new ways of farming, but farmers will have to decide whether and which technology is cost-efficient. Decisions about what works where, when, and how need to be taken in the field. Scientists need the cooperation with farming practitioners who share their knowledge and experience. What seems like a good idea from lab research may not work on the farm, and farmers face practical problems that need scientific understanding before solutions can be found. Alfred Grand is one of the farmers who are collaborating closely with researchers.

Alfred Grand in his worm compost 'factory'

Farming with benefits

Absdorf is a market town on the river Danube in Lower Austria, some 30 kilometers northwest of Vienna. It's a fertile region, not just for arable, but for sugar beets, fruit and vegetable growing and vineyards. Grand manages a 225-acre farm which has been fully certified organic since 2006. The farm has been no-till for over 25 years. For Grand, the wellbeing of his crops is as important as that of all soil life. He takes care of earthworms like other people look after their pets. Soil biology is also a decisive factor in planning rotations. Two years in a row, Grand grows alfalfa as it's an excellent fodder crop and supplies the soil with nitrate – which benefits the next crop: wheat. Cover crops follow after the wheat harvest in July. Grand uses a mix of ten to twelve different species, each of which has a job to do: mustard suppresses soil pathogens, buckwheat makes phosphorus more accessible to plants, tap roots help aeration and water infiltration. Some species are winter hardy. Others will be killed off by the first hard frost and form a layer of organic matter on the surface. As long as the weather stays cold, the soil organisms remain inactive. They can't decompose the vegetative matter and with a lack of oxygen, toxic, anaerobe processes set in. But winter-hardy cover crops like rye and winter vetch start growing again from as early as January. Frost has broken up and loosened the soil, and the roots can push down. The following May, Grand plants corn, which is followed by hemp[6] in year five of the rotation. Next, more cover crops are planted, again a combination of various species but this time nitrogen-fixing legumes will not be in the mix. The cover crops precede soybeans, which can fix nitrogen too. If he planted legumes before a soybean crop, the beans would 'get lazy', says Grand. Whether he then plants more cover crops after the harvest or starts a new rotation cycle with alfalfa depends on the condition of a particular field, the weather, and a soil analysis.

For Alfred Grand, farming was not a childhood dream come true. 'I was a really bad student and had no idea what I wanted to do with my life,' he says. The decision to attend a college for viniculture had less to do with his interest in vineyards than his father's promise of a motorcycle if he graduated successfully. Neither of them had any inkling that during the course, Grand would discover his true passion: earthworms. A series of lectures on compost kicked things

---

6   US farmers too regard hemp as a cash crop that is gaining in importance. Hemp is grown illegally for THC, the main psychoactive cannabinoid. But the health benefits of another cannabinoid, CBD, are increasingly being recognized. Hemp fibers can be used for pretty much anything, from cloths for clothing to building materials and even road surfacing. Hemp is a robust plant, not prone to many diseases, and the plants have long tap roots, which is good for the soil structure and bad for competing weeds.

off. At the time, the Austrian government made sizeable grants available for compost projects. Alfred Grand reckoned that compost could be used to produce energy and heat. He started to look online for 'hot compost' and found numerous references to worm compost. In the course of his research, he visited scientists at the University of California, Berkeley, who had valuable data but no plan for making practical use of their findings. With that in mind, Grand went back to his farm and decided to intensively raise earthworms and produce worm compost.

The worms reside in a huge polytunnel on the far side of the farmyard. It's covered by green sheeting, which protects not just against rain, but sun, too. Inside are a number of flat troughs, each 2.40m long and with a permeable bottom. There are two stages to worm compost production. The first phase is the production of feed for the worms. Lucerne is mixed with horse manure and composted. The hot rotting phase takes between four and six weeks during which the compost pile has to be turned several times. Finding the moment at which the compost has matured into perfect worm feed isn't easy: if the mix still contains too much raw organic matter, heat develops, and the worms die. If it's been left to mature too long, there isn't enough biomass left for the worms to feed on and they starve. The worms take about ten weeks to 'process' this feed mix. What's left drops through the bottom of the troughs in the wormery and accumulates in steadily growing piles of finished worm compost. It just needs to be sifted and packed.

Worm compost contains microorganisms and 'good' bacteria which boost soil life; the same process happens in nature, but it takes a lot longer. Over the years, Alfred Grand has become an earthworm expert: he achieves by hot rotting the compost what microorganisms and worms do in the field. The species living on the soil surface digest organic matter such as the mulch layer left after mowing cover crops. Other worm species build vertical tunnels through which they drag partly decomposed organic matter deep into the soil. In the process, they produce worm castings which become the inside cladding of the tunnels. This renders them more stable, and helps water infiltration. Deep in the soil live yet other worm species which build horizontal tunnels, leaving fertile worm castings behind.

By applying concentrated worm castings or vermicompost, gardeners and growers can boost soil life enormously. But what works in a vegetable patch, raised beds or a greenhouse cannot be scaled up to field level: it would be impossible to produce the necessary amounts of worm compost[7], and spreading it would require a lot diesel.

Other farmers work with compost tea. Sophie Alexander is an organic cereal grower in the South of England and participates in a research project by the Organic Research Centre[8] in Gloucestershire. For three years, she applied compost tea three times during each growing season. Eight thousand liters of compost tea were produced from 100 kg of compost – enough for 100 acres (40 ha). Compost tea is produced in huge vats where a mix of compost, water and a catalyst like seaweed is constantly turned and aerated in a 24-hour period. It's a laborious and costly process for rather disappointing results, says Alexander.

There are other issues with compost tea, says Alfred Grand. When it's applied to the field, it not only boosts the planted crop but the weeds, too. Grand has solved the problem by treating crop seeds with worm compost. Compost tea made from worm castings contains a wealth of nutrients, soil bacteria, phytohormones and humic acid. 'That's like a starter culture for a seed,' says Grand. His method of seed coating is still a bit improvised: seeds and worm compost tea are fed into a concrete mixer and turned until all liquid has been absorbed. Then the seeds are spread out for drying, the liquid quickly evaporates, and what's left is a layer of nutrients. 'The seed is coated with bacteria which will give the germinating seed immediate access to nutrients or make them accessible to the roots,' says Grand. In future, he will add clay dust to the compost tea, which will further increase the microorganism content. 'A seed needs warmth, water and nutrients to germinate. Germination will only happen when the conditions are right. By adding clay dust to the compost tea coating, we can extend the period of optimal growing conditions,' says Grand. 'The goal is to inject diversity into the soil and thereby stabilize the soil system.'

Worm compost tea is to soil what food supplements are for us, says Grand. In good soil not a lot will happen, but in degraded soil changes will happen fast, and improvements will soon be visible. Compost tea made from worm castings has another advantage: one doesn't need a lot, one liter per hectare is sufficient.

---

7   There is not enough biomass available on a farm to produce that much compost. In many towns and cities, green waste from private households is collected separately and composted with waste from public parks and green spaces. But even if one ignores the fact that such compost is often contaminated with plastic, the volumes are still far too small for what is needed at farm level. By selling straw to a mushroom grower and buying the used bedding back as fertilizer, Stephen Briggs has found an ingenious way of exporting biomass and reimporting it as compost. But not every farmer farms next to a mushroom grower.
8   http://www.efrc.com/

In comparison: Sophie Alexander worked with 250 liters of (ordinary, non-worm) compost tea per hectare. Worm compost tea has a lot of potential, says Grand, as components could be added that would help fight predators and diseases. He gives an example: there is a fungus that feeds off chitin, a substance that is found in the exoskeleton of beetles. In conventional agriculture pesticides would be used to destroy beetles considered to be a pest. 'Alternatively, you could promote the growth of fungi; ground up shrimp waste contains a lot of chitin. Once you add that, the fungi will multiply until they are so numerous that they start attacking the beetles,' says Grand. 'By providing a better habitat[9], you enhance growth and procreation. With everything you do you have to think: what is my goal, do I want to have aphids or roses?'

'We still know far too little about the interaction between plants, microorganisms, and fungi,' says Grand. 'Soil still is a black box'. Through tillage and other methods of soil cultivation farmers constantly interfere with the soil system. Via Twitter, Alfred Grand (@VERMIGRAND) got involved in the international discussion between farmers and scientists about compost, earthworms, soil microbes, and no-till agriculture. Once he 'outed' himself as a no-till organic farmer, there was quite some pushback: organic farmers in the US took umbrage with his frank criticism of tillage, and conventional farmers were enraged that Grand dared to think aloud about an agricultural future without glyphosate. But the Twitter sphere also got him in touch with the Rodale Institute in Pennsylvania and its executive director, Jeff Moyer. In 2017, Grand organized a conference on

---

9   Alfred Grand isn't the only one looking for opportunities to create a habitat that the crops love while making life for pathogens and pests as hard as possible. The Iyarpadi Tea Estate is situated at the heart of the Anamalai Mountains in South India. Dr S. Marimuthy runs the plantation's laboratory and is in charge of field trials. The organic fertilizers that are used in the tea garden have to be locally adapted, Dr Marimuthy told us when we visited the Tea Estate. He and his team take virgin soil samples from nearby forests and identify beneficial fungi. In the lab they test whether these fungi can be established on soil samples from the tea estate and if that's the case, the mycorrhizal fungi are bred in large numbers and applied to the tea bushes as fertilizer. He doesn't fear competition from agrochemical companies such as Bayer-Monsanto, because everything they produce should work anywhere in the world. In his opinion, the soil system works best when soil life is diverse and in tune with the local conditions.
Marimuthy uses a similar technique do develop organic pesticides. Heliopeltis is a sucking insect that attacks the tealeaves. The team did a number of tests and found that there is a fungus that attacks heliopeltis. Infested tea bushes are now sprayed with a solution containing lab-grown fungal spores. Beneficial insects such as ladybirds and pheromone traps usually finish the pest off. Using this approach, the scientists found ways to fight other pests, such as red spider mites or short hole borers (euwallacea): they use plant extracts to create an inhospitable environment for pest species while at the same time improving the habitat for beneficial insects, bacteria and fungi, thereby increasing their number.

organic agriculture. The conference venue was his farm, where 40 farmers from 15 countries attended and Moyer gave the keynote address. A month later, Alfred Grand flew to the US to discuss a possible cooperation. That hasn't happened yet, but several other projects are going full steam ahead. Agroforestry is one of them, the other is the 'Grand Garden'. Together with university scientists, Grand is developing a prototype of a market garden: organic fruit vegetables and flowers, grown intensively on relatively small acreages and sold locally. The gardens can vary in size, from two to seven acres. Vegetables are not grown as row crops, but in permanent beds which are accessible from two sides and can be planted densely. With a well-planned crop rotation, exact timings for seeding, and planting plugs, between two and five harvests can be achieved. Because the crops are chosen so that they complement each other in their needs, and rather than depleting the soil, biodiversity and soil fertility are maintained or improved. The up-front costs for starting such a market garden are low and therefore provide an opportunity for young people to get started in agriculture. All produce would be sold regionally. And as always, Grand dreams big: if it were up to him, every midsized town would have one or more market gardens by 2035, supplying customers with fresh, organic, locally produced food.

"We've maybe got ten years"

In the Midwest and on the High Plains, in California's water-starved Central Valley and in tropical Hawai'i – by using regenerative agricultural practices farmers and ranchers have found ways to make agriculture sustainable and 'climate friendly'. For all the farmers and ranchers we visited, regenerative agriculture had been something new that they needed to discover and explore. Each of them has become an expert – by researching and reading, by observing what's happening on their farms, by understanding interconnectedness, by reaching out to other regenerative farmers, often at the annual No-Till on the Plains conference and through the help of a small network of scientists who give lectures, advise and quickly provide an answer to an urgent question by email and out of hours, if that's what it takes. But in the end, what counts is practical farming and the ability of farmers and ranchers to solve problems on their farms and on their own.

Successful regenerative farmers apply a trial-and-error approach, while minimizing their risks. It takes courage to relinquish the farm bill-assured safety net of subsidies, crop insurance and disaster assistance programs. It takes real guts to farm without agrochemicals, without synthetic fertilizers, and no or only very occasional use of herbicides. Instead, these farmers and ranchers rely on observation, experience, and knowledge to farm with nature rather than against it. For some of the farmers we visited, it wasn't just courage, but courage born of desperation that made them rethink and change course. Having nothing left to lose, at times, may help to mobilize energies one didn't know one had. However, it isn't a good approach to changing the system of industrial agriculture from the bottom up or motivating farmers everywhere to work sustainably, using regenerative and organic farming practices.

Up to now, the number of regenerative and organic farmers and ranches in the US has stayed extremely small. They are tiny oases in a desert of industrial agriculture. On their land, farmers and ranchers can initiate real change, they can be pioneers, and they can motivate and train others. But that's not enough to combat the climate crisis, to create food security and change the food system. For that, a fundamental change in agricultural policy and practice is needed. Such change is impossible without political support. Scientists have to support the practical work of regenerative and organic farmers through research, as well as develop guidelines that help others to start conversion or to work more efficiently.

Farming with benefits

The State of California is already leading the way with comparatively rigorous environmental standards. To better combat the climate crisis, the state government is now working with scientists and farmers: the Healthy Soils Initiative incentivizes the use of regenerative agricultural practices – which, as a 'side effect', lead to carbon being stored in the soil. As part of the program, several universities work with farmers and ranchers to establish which farming methods are most effective and under which conditions.

Kate Scow with the planting plan for the University Davis, Russel ranch test plots

Davis is a city just west of California's capital, Sacramento. And it is home to UC Davis, one of the most renowned agricultural universities in the United States. Kate Scow is professor of soil science and microbial ecology and director of the Russell Ranch. The 300-acre Russell Ranch is just outside of Davis, and is where we have agreed to meet on a warm November day. The ranch is part of the university's department of land, air, and water resources and 'the research cross section between academia, policymakers, and farmers,' says Scow. 'Thirty years ago, the faculty decided that long term farm research was needed, and we are now[1] 25 years into a 100-year study.' One of the main goals is to establish indicators for sustainability in agriculture, and that means looking at ecology as well as economics and soil health, among other parameters. As Kate Scow starts

---

1 2018

to describe the trial set-ups and their time lines, it becomes clear how much time and effort it takes to scientifically test the difference in efficiency of different farming practices.

The long-term trial is run on 72 one acre-sized plots. In order to get reliable results, each experiment is repeated three times in each phase; to test a three year crop rotation, for example, nine plots are needed.

Two main experiments run long-term: in the first, wheat is grown in one year and the plot is fallowed the next, no fertilizer is applied, the plot is not irrigated and no cover crops are planted. The second experiment is a corn tomato rotation which compares different agricultural practices. For a certified organic version of the corn tomato rotation, certified organic chicken manure and compost are applied as fertilizers, and cover crops are planted over winter. The conventional approach uses none of the above, while a mixed version works with cover crops only. Seven years ago, alfalfa was added to the rotation. All plots are irrigated.

Other, shorter-term trials focus on soil health. In a tomato trial, a comparison is made between compost (garden clippings), chicken manure or the application of compost with added microbial feed. The preliminary result: compost and poultry manure work well. In the next phase, Scow and her team will test whether cover crops will bring an additional benefit. All experiments are designed to produce data that are relevant for farmers and agriculture in California: irrigation for winter wheat, use of manure as fertilizer in a corn wheat rotation, possible advantages of using biochar or inducing water stress in tomato crops prior to harvest. Many farmers in the hot but very dry southern part of the Central Valley grow tomatoes for canning. Growers could save water if they ended irrigation some time before the harvest. The ripe tomatoes will contain less water, have a more intense flavor, and be easier to process.

In future, livestock will be included in the trials. Many farmers plant cover crops, but use herbicides to 'terminate' them before seeding cash crops. As we saw on our farm visits, environmentally and economically, it makes much more sense to use cover crops as feed. At the Russell Ranch, the benefits will soon be assessed scientifically.

In Kate Scow's opinion, microorganisms are central to soil health and fertility: 'Microbes are the eye of the needle through which carbon flows. Inputs like manure or plant residue are turned into stable soil organic matter.' But where, in which soil layer, is organic matter stored? Is it possible to speed up the storage process? The experiments at the Russell Ranch now start to yield results. 'We find that when we combine composting, cover crops and organic farming practices, stable carbon will be stored at a depth of two meters,' says Scow. In the upper soil layer, carbon might be more vulnerable. 'We need to get carbon lower into the ground. People have to look deeper into the soil if we want to achieve carbon sequestration.'

When we visited the Russell Ranch in November of 2018, Kate Scow and her team were just in the process of evaluating the findings of a nineteen-year experiment with a corn, tomato, wheat, fallow rotation managed with different farming practices. One of the goals was to compare the amount of soil organic matter at a depth of two meters. The results[2] were published in August 2019: 'Conventional soils neither release nor store much carbon. Cover cropping conventional soils, while increasing carbon in the surface 12 inches, can actually lose significant amounts of carbon below that depth. When both compost and cover-crops were added in the organic-certified system, soil carbon content increased 12.6 percent over the length of the study, or about 0.07 percent annually. That's more than the international "4 per 1000" initiative, which calls for an increase of 0.04 percent of soil carbon per year. It is also far more carbon stored than would be calculated if only the surface layer was measured.' The decisive factor might be whether compost is worked into the soil or not. 'One reason we keep losing organic matter from soils is that our focus is on feeding the plant, and we forget the needs of others who provide important services in soil like building organic carbon,' Kate Scow was quoted in the UC Davis press release. 'We need to feed the soil, too.' And the soil diet needs to be well balanced: if soil organisms aren't fed properly, they will draw nutrients from soil organic substances already stored in the soil. If that happens, no new carbon will be stored and existing carbon stores will be depleted. Scow and her team assume that cover crop roots deliver enough carbon to the soil organisms, but they lack the nutrients they would need to stabilize the carbon.

And the experiment at the Russell Ranch produced another surprising result: much of California has a semi-arid Mediterranean climate, and under such conditions, adding compost to the soil will lead to a significant increase in carbon sequestration.

How much compost is needed for soil improvement, and is it even possible to produce large enough volumes? Questions Kate Scow answered by email: Per one acre plots four tons are needed, about 10 tons per hectare. 'Producing such high quantities will take a lot of creative solutions matching sources of organic waste with needs for compost (i.e. dairy operations linking up with row crop

---

2 https://climatechange.ucdavis.edu/news/compost-key-to-sequestering-carbon-in-the-soil/

farmers in the Central Valley; urban composting facilities serving urban farms and other farms in close proximity; etc.) [...] We believe that most of the compost effect is coming from the substantially increased carbon inputs that the compost provides, and there also seems to be evidence that by balancing nutrient ratios of carbon, nitrogen, phosphorus and sulfur (C:N:P:S) microbes are more able to build biomass (which becomes future SOM [soil organic matter] and less likely to consume old SOM.' Even at state level, the importance of compost is being recognized. 'The State of California is activity working to increase compost production in the state with incentives,' Scow wrote in her email.

California is the first US state which recognizes regenerative agriculture as part of the climate change solution because it can help carbon sequestration. Jeanne Merrill is policy director at the California Climate Agriculture Network, CalCan, for short. CalCan is a coalition of farmers, farm groups and others, aiming to translate the framework of government policies into practical programs promoting sustainable and organic agriculture.

When we arrive at the modest office suite in downtown Sacramento, Merrill is on the phone – California will soon have a new governor and the race is on to secure funds and make sure existing programs will not be shut down. As we wait, I feel somewhat uneasy. I try to do my homework before an interview, and usually come prepared with a list of questions, but in this instance, I got lost in the maze of laws, initiatives, proposals, all packed with abbreviations. Luckily, Jeanne Merrill agrees to start with a short summary and crash course in state environmental policies. In 2014, California introduced cap and trade auctions of permits to force industries to either reduce their CO2 emissions or pay to offset them. At the same time, the decision was made to invest part of the proceeds into regenerative and 'climate smart' agriculture. One of the first programs was SWEEP, the State Water Efficiency and Enhancement Program, which incentivises measures aimed at saving water and using it more efficiently. Next came AMMP, the Alternative Manure Management Program. It provides financial assistance for composting and other environmentally friendly ways to deal with the enormous amounts of (liquid) manure produced in California's many dairies, among them mega dairies with several thousand animals in one place. The 'Healthy Soils' program came last. It promotes farming practices that increase soil quality because healthy soils contain more soil organic matter – carbon sequestration is an indirect effect. The California Department of Food and Agriculture, CDFA, decides how the available money is distributed between programs. Farmers and ranchers don't receive direct payments but have to apply for grants and loans for new projects and the continuation of existing ones. The paperwork for the application is so complex that most farmers work with an agronomist or extension specialist to complete it. Nevertheless, since the 2016/2017 budget, when the program started, the number of applicants far outstripped the available funding. 'Until we significantly scale this up it will not be a game changer, but it certainly has the potential to be one,' says Jeanne Merrill. The political will is there: In 2018, the government announced that by 2030, a million acres should be under the 'Healthy Soils' scheme, allocating 50 million dollars per year. CalCan works on doubling both, the number of acres and the funds, to two million acres and 100,000 million dollars annually. Merrill estimates that in California, 600,000 acres are already under an organic or regenerative agricultural system. (There are roughly 100 million acres of land in California, with 43 million acres used for agriculture. Of these, 16 million acres are grazing land and 27 million cropland[3].)

In California, efforts to curb CO2 emissions and set legally binding targets began in 2006. A bill by the California Assembly (AB 32) set the target of reducing GHG by 20 percent under the 1990 level by 2020. This goal has already been reached, says Merrill, and a 2016 California Senate Bill (SB 32) now stipulates a 40 percent GHG reduction by 2030. And in the fall of 2018, the outgoing governor, Jerry Brown, issued an executive order saying California should be carbon neutral by 2045. It's an ambitious goal which still has to be made into law. Jeanne Merrill is not sure whether it will be achievable, but she is certain that regenerative agriculture will be an indispensable part of any solution. A first step would be to integrate all existing programs financed through proceeds from the cap and trade auctions. The UC Davis Russell Ranch study provides scientific proof that applying compost is one of the best methods to sequester carbon – a close cooperation between dairy farms in the Alternative Manure Management Program, AMMP, and farmers in the 'Healthy Soils' Program could achieve real synergies. But whatever the next steps, more and better funding is needed, says Merrill. In her opinion, companies should pay increasingly more to offset emissions.

In the United States, California is leading the way in regard to strict environmental norms for climate change mitigation. If regenerative agriculture is seen as part of the solution, why then are farmers not paid directly for carbon sequestration, and are instead incentivised through the 'Healthy Soils' program to switch to different practices? As our conversations with farmers and soil scientists have shown: it's complicated... Even if all available regenerative farming methods were consistently applied, the increase in soil organic matter would not be linear.

---

3   https://www.cdfa.ca.gov/agvision/docs/Agricultural_Loss_and_Conservation.pdf

How much stable carbon can be stored depends on a whole host of factors: geographical region, micro climate, soil type, to name but a few. Projections for how much carbon can be sequestered through regenerative agriculture are estimates at best and blind guesses at worst. California has avoided potential rows over soil carbon measurements and how to financially reward farmers by incentivising environmentally friendly agricultural practices which increase the potential for carbon sequestration.

In the summer of 2019, Indigo Ag, an agtech company headquartered in Boston, launched a worldwide carbon project. The 'Terraton Initiative'[4] guarantees farmers US$15[5] for each ton of carbon sequestered in the soil through regenerative agriculture or tree planting. Indigo Ag offers seed treatments that are supposed to stimulate soil organisms as well as technological and logistical support. According to a 2019 Financial Times article[6], the company received $650 million in funding. 'Indigo said it hoped to sign up more than 3,000 growers, covering more than one million acres this year. […] It plans to sell the carbon credits that can offset a company's inherent emissions to the food and agriculture sector. […] Indigo says it will use satellites and image analysis to measure soil carbon sequestration and on-farm emissions.'

In July 2019, Jeanne Merrill weighed in on Indigo Ag in a piece for the US online food and ag publication Civil Eats. She described not just the imponderability of trading carbon credits but also reminded readers of the collapse of the Chicago Climate Exchange in 2010. Too many farmers had tried to participate in the scheme. 'The carbon market was swamped with offset credits from willing farmers, but it didn't have enough buyers. As a result, the price of the carbon credits went from a high of roughly $7 to just 5 cents per metric ton of carbon', Merrill wrote. California has taken a different approach which is proving to be successful even though no direct payments are made to farmers. 'The Healthy Soils program has seen farmer interest grow despite concerns from some that payment rates – what farmers will receive for new practices – are too low and the application too complex. The average Healthy Soils grant is a little over $56,000 for three years. Total greenhouse gas emissions reductions associated with the program are nearly 40,000 metric tons of carbon dioxide equivalent, which is the equivalent of taking 8,400 cars off the road'. And Merrill raises another important point: the microbial seed treatment Indigo Ag offers to farmers. 'Farmers can opt to use the seed coatings and they'll also have an additional layer of help from a team of

---

4   https://www.indigoag.com/terraton
5   There are few details yet as to the time frame and how carbon sequestration would be measured.
6   https://www.ft.com/content/83c1da2a-8c70-11e9-a1c1-51bf8f989972

agronomists and the latest in Big Data analysis to support their participation in the Indigo Ag carbon initiative. This potential layering sets the Indigo Ag approach apart. It views its entry into the soil microbe/seed treatment business and the carbon market as a disrupter. But who will benefit? Will the company scale up with the intention of being bought out by a larger ag or tech industry player – much like Climate Corp sold to Monsanto and Blue River sold to John Deere?' In the past few years, almost all big ag companies have bought up small tech companies, in particular companies specializing in data analysis. It seems as if farm data is the new 'big thing' in agriculture. Many companies secure their right to accessing this data in the standard sales contracts farmers have to sign with the purchase of new equipment. With every field pass, the board computer of a tractor or combine generates valuable data that may well help farmers to take better decisions, but the manufacturers usually make sure that they, too, can access and use such data. Mostly, the contracts do not specify how the data will be used and whether and in which form it may be sold on. Indigo Ag's 'Terraton Initiative' for carbon sequestration may just be a sideshow. The company offers other services, such as Indigo Marketplace, a platform farmers can use to sell produce and companies to source it. US ag journalist Charlie Mitchell did a deep dive into what Indigo Ag is all about and came to some worrying conclusions: 'The idea isn't just to monetize data—all technology companies want to do that. Indigo's goal is instead to create an agriculture company that operates more like Google or Facebook, companies with information resources so vast that their services go unopposed. Why? Because to do what Google does as well as Google, you'd have to have its data resources. And you don't. […] Indigo's potential monopoly isn't just consumer-facing: It also could look back up through the other end of the supply chain. Like Google, Amazon, and Apple, Indigo wants to lock users—in its case, farmers—into a product ecosystem that is too powerful and convenient not to use. The potential implications of this are worth considering.'[7]

In comparison, the 'Healthy Soils' initiative takes a simple and straightforward approach: it incentivizes farmers to switch to environmentally friendly practices and regenerative farming methods. Carbon sequestration is a beneficial side effect. Sean McNamara, the organic walnut farmer we visited in Winters, west of Sacramento, applied for a 'Heathy Soils' grant in 2016, the year the initiative started. A small section of land had been leased out. It had just been brought back into the farm and planted with walnuts. McNamara was able to get a grant for composting and cover crop seeds because it was a new project. He would love to get a grant to buy a better compost turner which would improve the quality

---

7 https://thecounter.org/indigo-agriculture-disruptive-startup/

of the compost. Even though compost production is one of the very efficient ways to sequester carbon, McNamara is not eligible for a 'Healthy Soils' grant because he is already composting. Sierra Orchards has been certified organic for decades, and the McNamaras are already doing everything they can to improve soil health and mitigate climate change – which excludes the orchard from the grant scheme. Apart from the fact that the application process is very complicated, it disadvantages early adopters of resilient agricultural practices, says McNamara.

It seems like a valid criticism to me. Why are those who had the courage to go down new paths excluded from grants? A question I put to Don Cameron. He is the vice president and general manager of the 6,000 acre (2,400ha) Terranova Ranch, Inc. on the west side of the San Joaquin Valley. He is also the president of the California State Board of Food and Agriculture. The board deals with agricultural, environmental and sustainability issues and advises the state's governor and the Secretary of Agriculture. A few years ago I had visited Cameron at his farm because he had developed a method to flood orchards during the rain-rich winter months, which helps to recharge the groundwater – a very important contribution to farming sustainability in the water-deprived southern part of the Central Valley. But this time, I wanted to get the take of the advisory board – why are farmers who have been successfully using organic and regenerative practices for years excluded from the funding? Cameron says programs such as 'Healthy Soils' aim to motivate as many farmers as possible to rethink the way they farm and try out new practices. This is not about the farm pioneers at the head of the organic and regenerative movement, it is about broadening the base. 'We have to bring consensus into agriculture. We have to bring everyone into the fold, and for that we have to hang out a carrot'. That may not be enough for the big growers in the Central Valley and the application procedure is complicated, he admits, but 'we are listening, we are making changes. If you want to be effective you have to have everyone on board and expand.' He would like to spend more money on information and education, and farmers are to get more help dealing with the application process. 'If you have the money, the acres will follow'.

The 'Healthy Soils' scheme is relatively new, and the criteria are still being amended, Cameron says. Carbon sequestration in the soil remains the main goal. On his farm, Cameron has been working with compost for over 25 years, but that's still an exception rather than the norm. 'Healthy Soils' incentivizes compost production, working with cover crops, planting hedges and flower strips to create a good habitat for pollinators and beneficial insects. Cameron says money is needed for research to establish which agricultural practices are most efficient and beneficial – for both the environment and agriculture. Such practices should result in healthier plants, better yields, or more efficient water use, and should translate into a financial benefit for the farmer. Research and science should give farmers a good guideline as to which practices are likely to work on their farm. Don Cameron has seeded cover crops on his farm, but the western part of the San Joaquin Valley is extremely dry, and the seeds needed to be irrigated or they would not germinate. In his opinion, the cost of water far outweighs the soil benefits cover crops would bring, which makes them an unsuitable choice for agriculture in very dry regions.

Cameron hopes that major food companies will come on board. California produces a lot of fruit, vegetables, and salad. At present, 'they are only concerned about food safety and price,' says Cameron. 'We need to get them interested in the sustainability of agricultural production methods.'

There are three to four consultation rounds per year. At the California State Board of Food and Agriculture, one keeps an open mind; there is some 'flaky stuff', as Cameron puts it, but there are also some great ideas which can be developed further and could become policy. 'We started from zero and it's been a long way to where we are now in a short period of time.'

Our conversation takes place at the end of our trip and two days after the 'Camp Fire' broke out. Over Chico, the sky is still dark, big flakes of ash drift down silently, covering everything with a thin, whitish blanket. To the northwest of Los Angeles, high winds are fanning the 'Woolsey Fire', it's spreading fast, an evacuation order has been issued, and tens of thousands of people have to leave their homes. I ask Don Cameron whether the fires are noticeable in the San Joaquin Valley. Not yet, he says, but that may just be a matter of time. He tells me about another fire which broke out near Redding in the north of California a few months earlier. The 'Carr Fire' burnt for five weeks and could be felt in the southern Central Valley, some 350 miles away. At times, thick smoke lay over the valley like a dark blanket and reduced the light intensity of the sun by 15 percent, and cooled the air somewhat. The effect on plant growth was actually positive, says Cameron, as peppers and tomatoes suffered less sunburn damage. On some days, the smoke was so close to the ground that it seemed almost like fog, and the air quality was atrocious. Cameron has no doubt that both the intensity and the frequency of wildfires has increased. There have always been wildfires in California, and, when locally contained, they are one of nature's ways of rejuvenating forests: at ground level, fire burns through dead undergrowth and wood. Living trees suffer little damage, as the crown stays intact and the heat releases the seeds from the cones. Now there are more devastating fires, says

Cameron; they burn so hot that even the crowns of the trees are reduced to ash. This type of fire completely destroys the whole ecosystem.

Chico University Farm at noon on day two of the Camp Fire

The day before, wearing pullovers and fleece jackets, we had been sitting with Cynthia Daley in the office at the Chico University Farm. And of course we are talking about the 'Camp Fire' which, at that point, has been raging for 24 hours. Daley says that at the university, all lectures have been cancelled for the foreseeable future and the buildings will be used as an emergency shelter for people fleeing the fire. At the farm, it is eerily quiet. Thick, black smoke hides the sun, the little daylight that gets through has a strange red hue, familiar objects now have an unreal quality, and it feels as if the world around us has suddenly become a vast darkroom. Only once we are sitting in the small, crowded farm office with the lights on does the feeling of trepidation begin to lift somewhat.

Not just in California, the consequences of the climate crisis can neither be denied nor ignored, says Daley. The fire season has become much longer with more frequent fires that burn with higher intensity, droughts last longer, accompanied by extended periods of heat in which temperatures remain above 100F, the winter snowfall in the mountains has turned to rain, rainfall has turned into cloudbursts, and in cities like Los Angeles, dried out canals become torrential rivers and cause flooding. 'There is no normal anymore, all we have is abnormal,' says Daley.

'Maybe we have ten years to turn this around. We need all hands on deck to do this. We have to elevate consciousness to what is happening.'

Cynthia Daley: 'We maybe have 10 years to fix this'

In the face of the global climate crisis, Chico State University founded the Center for Regenerative Agriculture and Resilient Systems, CRARS. Cynthia Daley is the director and co-founder. 'We have exhausted our soils, we need to focus on carbon and build soil rather than mine it'. The agrichemical industry works hard at developing products that are supposed to increase soil biology and soil quality. Daley believes that the 'bugs in a jug' approach is as misguided as the supposedly 'fast and easy' solutions the industry has come up with so far: chemical fertilizers, herbicides, pesticides and additives. 'It's communities of biology that are essential, and it's the consensus building among those communities that really provides us with the juice, the engine that revs in order to get this system humming without all the inputs.' And the emphasis is starting to shift: what counts are not only the quantity and diversity of soil organisms, but also their ratio – are fungal and bacterial communities in balance? And is there enough food? There are a number of ways to enhance soil biology, but 'the soil organic matter in most soils is so low, it cannot support biology. It's not only about having the right biology, it's also about having the right groceries in the soil to feed that biology.' Daley works closely with David Johnson, a soil scientist at the University of New Mexico. Compost is rich in soil bacteria, but Johnson has developed a composting method that increases the fungi ratio. To avoid anaerobic processes, compost needs to be turned repeatedly. The effect is similar to plowing a field: the mycorrhizal structure gets destroyed. Johnson uses big barrels for compost making which are aerated from within through pipes. The compost matures for twelve months, much longer than 'ordinary' compost, and that gives the mycorrhizal systems plenty of time to grow. It's the fungi to soil microbe ratio that matters – if the balance is right, it boosts plant growth, and more carbon can be sequestered[8] than even in no-till systems. It's not yet possible to sufficiently scale up Johnson's composting method, says Daley, but his work has been very important to understand soil fertility. Field trials have shown that a 1:1 ratio of fungi to soil bacteria or even a slight predominance of fungi is ideal. In the lab, scientists are now trying to find out why that is the case. Metagenomics, the DNA sequencing of soil samples from test fields and their comparison, should help to provide some answers.

The pioneers of regenerative agriculture have tried and tested different farming methods. 'Scientists need to go to the farms and study why these systems work so well', says Daley. How well something works is measured against 'key performance indicators' or KPIs. Among them are water infiltration rate, total amount of soil carbon, tilth, aggregate stability, total microbial biomass, and total respiratory rate (which also is an indication of biomass in the soil) and changes over time. Once the baseline for a farm has been established, a carbon plan can be developed that will specify how soil quality and fertility can be improved further.

By now, some basic principles have been well established. 'Soil should never be bare, and there should always be a living root,' says Daley. She does not accept the argument farmers in areas such as Kansas make, who say that there just isn't enough moisture to grow cover crops. She cites Ademir Caligari, a Brazilian agronomist who worked for the UN and showed that two inches of rain per year are sufficient to grow cover crops. It's an investment for the future because better soil quality is the pre-condition to long-term solutions. Diversity – in plants, in soil biology, in fungi – is key to success: 'The diversity of those root tips providing exudates from diverse plants provide groceries for a very diverse population of microbes and fungi and in that process you really build your soil organic matter in a huge way.'

---

8   https://www.csuchico.edu/regenerativeagriculture/bioreactor/david-johnson.shtml

Cover crops, incorporating green manure into the top soil layer, composting, mulching, going no-till, or plowing very rarely, long rotations, integration of livestock – all of the above are tried and tested practices in regenerative agriculture which help to improve soil quality. No single practice is a silver bullet; everything depends on a farmer or rancher choosing the right methods for his operation. Agricultural machinery is improved, new tools are being developed: the design of a rotor tiller used to prepare a seedbed matters, for certain shapes work better in some soils than in others where they can create a plow pan, or parallel bars may be better. A tine harrow does not just deal with weeds but can also stimulate plant growth. 'Once three percent soil organic matter has been reached the system starts to function properly and soil organic matter really starts to increase'.

The Chico State Regenerative Agriculture Initiative bases its research on cooperation between scientists and farmers, as 'they will chart the course'. The research happens in the field and not in the lab. Farmers decide on the research questions, scientists help to set up field trials and with data collection, and they provide analysis. Many of the farmer pioneers have become mentors to others, and Daley and her colleagues support this crucial part of the work: they organize field days, they showcase farms and farmers on the CRARS website and through social media, they shoot educational videos in the field, and they organize lectures and trainings. All mentors work with trainees and teach a new generation of farmers and ranchers how to work sustainably and profitably. Very slowly, the network of likeminded farmers who support each other is growing – it matters that those new to regenerative agriculture learn that the pioneers, too, struggled with problems early on and continue to be confronted with new ones, says Daley. The important point is how they dealt with them and what it took to solve them. 'Regenerative agriculture is an art more than a science,' she says.

As for practical farming, it's not only the goal that is well defined, but the path to get there is too. 'Science needs to support farmers. We need boots on the ground to effect a change in thinking. We KNOW what we need to do. We know the direction. We now need people to come on board and go down that path.'

Daley acknowledges that for many farmers, marketing their produce remains the biggest problem. 'If you are not economically sustainable you, are not sustainable at all.' She would like to start a conversation with food producers who only look at processors in regard to sustainability. They need to have a closer look at their supply chains and source more from regenerative and organic farms. Through their buying practices, big companies have control over a lot of acres, says Daley. Certification through independent auditors will be needed, and 'made with ingredients from regenerative agriculture' could become a good sales argument for food companies. But such a 'regenerative label' must have a cost. Companies have to pay better prices, as only then will more farmers change their practices and engage with regenerative agriculture.

Out on grass and living the good life

Will food just become more expensive? Yes, consumers may notice slight price increases at the supermarket checkout, but the additional costs are negligible compared to the benefits of regenerative agriculture: for the environment, in combatting the climate crisis, and for long-term, global food security. We are already seeing increasingly chaotic weather patterns oscillating between drought and flooding, pollinator numbers are decreasing, and pests and plant diseases are adapting fast and advancing. Under such circumstances, many farmers will have a hard time growing anything at all. In the UN climate report[9] scientists have spelt out clearly what the future will look like when crop failures become the norm rather than the exception. Even with the support of universities, farmers and ranchers can't do it alone: political will, financial incentives and legally binding regulations are needed to get processors and food corporations to reorganize supply chains, source ingredients from organic and regenerative agriculture, and market the products.

---

9   https://news.un.org/en/story/2019/08/1043921

And regenerative agriculture is not only important in combatting climate change. There are now clear indications that consumers will see immediate benefits because in an organic or regenerative system the quality of food improves. Numerous studies have shown that the nutrient density in fruit and vegetables is continuously decreasing. Today, we need to eat much more produce to ingest the same amount of minerals, vitamins and trace elements. The soil scientist Kris Nichols stresses the connection between good soil and higher levels of antioxidants such as flavanol, beta-carotene, lycopene, and vitamins A, C, and E. More research is needed to understand how soil and antioxidants are connected.

Cynthia Daley and the Regenerative Agriculture Initiative in Chico are working on it. The scientists have teamed up with the Bionutrient Food Association[10], an organization that analyzes nutrient density in food. Their goals is to understand the connection between farm management systems, soil health, the environment, and the nutrient density of fruits and vegetables. In cooperation with CRARS, they hope to collect enough data to determine which soil parameters promote the formation of which phytonutrients and under which conditions.

The loss of nutrients in fruits and vegetables coincides with a loss of soil quality and soil fertility – a plant can only absorb nutrients and trace elements that are present in the soil. That's why the nutrient density in produce from regenerative farming systems is higher. And the quality of the soil on which a crop is grown finds its expression in taste and aroma. Vintners, chefs, and gourmets have known that for a long time and talk of 'terroir'.

Eating produce from organic and regenerative farms is an all-round win. By buying and consuming nutrient-dense, aromatic, seasonal and – whenever possible – locally produced fruit and vegetables, we start a positive chain reaction: more farmers and ranchers will use farming methods that increase soil quality. That creates an environment in which soil bacteria and fungi begin to thrive, and more carbon will be sequestered from the air. Because the soil ecological system is balanced, access to available carbon will be utilized in building soil aggregates which in turn create soil structure. Increasing the amount of stable carbon in the soil counteracts global warming. Good soil mitigates the effects of climate change, and it creates the conditions farmers and ranchers need to continue producing food. Food security world-wide hinges on good soil. It is not an overstatement: soil quality will decide whether future generations are going to eat or starve.

Meat, dairy, and eggs should be on the menu – produced by animals exclusively kept on grass or as long as weather conditions permit. And the emphasis really

---

10 https://bionutrient.org/site/

is on grass-based systems. Industrial agriculture, with animals raised and kept in confinement, mega dairies, thousands of animals kept in overcrowded barns: none of this is acceptable for a variety of reasons. Animal welfare considerations and the damage this type of animal agriculture does to the environment are just two of them. And yes, we need to eat far fewer animal products; the present level of consumption of meat, dairy and eggs is only possible because of intensive animal farming, and it is unsustainable and morally wrong. The animal products we do eat should come from grass-based systems.

To eat meat or not, to be a flexitarian, a pescatarian, a vegetarian or shun all animal products and be a vegan is a personal choice. But in the US and in Britain, a growing number of activists believe that a vegan lifestyle is the only solution to the climate crisis. A small, militant minority in the US has targeted restaurants and shops for selling meat. In Britain, beef and dairy farmers have received death threats, with their addresses, including directions how to get there, published on social media, plus cattle have been 'liberated' from barns and paddocks. Such criminal acts have created fear and insecurity among farmers who are forced to invest in cameras and expensive security systems. But even in its non-militant form, veganism isn't the cure-all for the climate crisis. On the contrary.

The soil scientists and farmers we talked to all agreed that animals, in particular cattle, are an integral part of any organic or regenerative farming system. Grazing animals allow these systems to truly flourish. (Poultry and pigs can be kept on grass, too, and bring additional benefits.) When grazing is well managed and the grassland is neither under- nor overgrazed, more carbon will be sequestered than there would be if the grass were mowed. When cattle 'terminate' cover crops through grazing, no herbicides are needed. But no farmer or rancher can afford the luxury of keeping livestock as 'eco-friendly lawn mowers' or for the bucolic scene farm visitors enjoy. Only if we eat animal products – in small amounts – does regenerative agriculture become viable and make financial sense for farmers. For meat, eggs, and dairy products from regenerative farms, no trees will be felled in the Amazon, no virgin land will be plowed up, and no CO2 will be released from the soil – it's the opposite: more will be sequestered and stored as stable carbon.

And there is another reason that makes a vegan lifestyle into a climate threat rather than a means of saving the planet: if you do not use leather and wool, you need to replace them with other materials. In summer, cotton and hemp are good alternatives, but winter clothes will be made from artificial fibres. Both, artificial fibers and plastic, are petroleum-based products – and it is because of the industrial use of fossil fuels that we are now in this climate crisis!

'We have another ten years at best,' says Cynthia Daley. Switching to regenerative and organic agriculture has to happen fast and on a grand scale. It's only achievable if food processors, food manufacturers, suppliers, and supermarket chains get involved. Only if supply chains can be changed and products from regenerative agriculture have a reliable market, will more farmers and ranchers be prepared to give up industrial agriculture and convert to a sustainable, environmentally-friendly system. Over the years, consumer pressure and targeted campaigns by environmental groups in Europe and the US have shown that companies and corporations will produce whatever can be sold. Customers decide which products supermarkets will stock and which will be delisted. In Germany, pressure by consumer and environmental groups has led to GM-free milk (dairy cows are not fed GM corn or soy) being available in pretty much every shop and supermarket. 'In August of 2019, 60.5 percent of milk was produced in compliance with the "without genetic engineering" standard. Once 3.8 percent of organic milk is added, 64.2 percent of milk from German dairies was GM-free,' the German farming publication 'Unabhängige Bauernstimme' noted in September[11]. Because of consumer pressure, chickens, too, are increasingly being fed with GM-free feed, and animal welfare has become a topic that national governments as well as the EU have to grapple with. Christian Meyer, agricultural minister in the German Federal State of Lower Saxony between 2013 and 2017, demonstrated how much can be achieved if the political will for change is there. In terms of agriculture, Lower Saxony is for Germany what Iowa is for the US. No other state produces more hogs and chickens, with most of them in confinement. Meyer's first move was to enforce the law: farm operations either complied with environmental regulations and animal welfare standards or the farmers would be fined. He incentivized organic agriculture, and he is probably best known for the 'curly tail' initiative: farmers who delivered hogs to slaughter with intact tails were paid a premium. The logic behind the scheme is simple: tail biting in pigs is more likely in overcrowded conditions, when the animals are stressed. In many industrial farm operations, cutting off the tails at birth, 'tail docking', is seen as a preventative measure. But the cash premium was incentive enough for lots of farmers to raise hogs to a higher animal welfare standard. Customers loved the 'curly tail' initiative and didn't mind paying a few cents more for the meat.

Regenerative (and organic) agriculture needs a powerful lobby. At present such a support 'structure' is in its infancy – anti-GMO campaigns and organizations fighting for a ban on herbicides such as glyphosate or pesticides like neonicotinoids

---

11 https://www.bauernstimme.de/news/details/?tx_ttnews%5Btt_news%5D=2712&cHash=af13cff0a282f9e12453309ef2642272

will often mention regenerative agriculture as an afterthought. And 'regenerative agriculture' is an unwieldy term; anyone not involved in agriculture may find the concept hard to grasp or fail to understand why proponents of organic agriculture often hasten to point out differences. More information is needed – one reason for writing this book was to put regenerative agriculture into context. How can we effect change, away from industrial agriculture? Protests and campaigns do work, but not everyone wants to spend endless hours leafletting or standing in the street, holding up a banner. Small, everyday initiatives can be very effective too: approach your local supermarket or a food manufacturer and ask about where they source from and under what conditions the fruits, vegetables, meat, milk, or eggs that are used or sold were produced. For industrially managed farms, conversion to a sustainable system such as regenerative agriculture holds many risks. Will companies financially support farmers who try to wean themselves off herbicides and synthetic fertilizer? Ask for products made from regenerative agriculture produce, request that they be stocked. If such issues and questions are raised often enough, and from all sides, even multinational corporations will start to listen – it's good business to sell what customers want.

And each of us eats food, several times every day. It's easy to engage with growers on a farmer's market. Most will be happy to answer your questions. What farming method do they use? What do they do to enhance soil quality? Is the farm diversified? Many farms have their own website and market directly. Others sell from the farm gate, run a farm shop or supply one. In most areas it's possible to get seasonal, locally grown fruits and vegetables through a veg box scheme or CSA box. Often meat, dairy products and eggs can be added. In the UK, the debate over glyphosate use right before the wheat harvest has led to the 'Real Bread' campaign, where young, artisanal bakers have formed baking cooperatives. They not only take their time and use traditional methods to produce delicious sourdough bread, but some work with growers to rediscover old or find new grain varieties that are rich in nutrients and taste[12]. In the US, vegetarians train to be butchers – 'out of a deep love for animal life and respect for the environment'[13], as the New York Times reported in a long piece in 2019. Every day and at every meal we have the opportunity to use produce from regenerative or organic farms, produced, raised, and grown by farmers and ranchers who know about soil ecology and who care for microorganisms, earthworms, and mycorrhiza as diligently as they do for their cattle, their wheat plants, vegetables, or orchards. Like these farmers we

---

12 https://www.theguardian.com/food/2019/oct/10/flour-power-meet-the-bread-heads-baking-a-better-loaf
13 https://www.nytimes.com/2019/08/06/dining/butchers-meat-vegetarian-vegan.html

Farming with benefits

can know about the connection between the climate crisis, good soil and healthy food. 'Eating is a political act,'[14] said author and activist Michael Pollan. Through regenerative agriculture, every meal can be your daily contribution to the fight against the climate crisis and for the future of food and our lives on this planet.

---

14 https://www.cornucopia.org/2008/11/michael-pollan-eating-is-a-political-act/

Acknowledgement

My thanks and deep gratitude go to the farmers and ranchers who took time to show me how they work, who shared their ideas, their successes, failures, ongoing worries and their hopes

As I write this, wild fires in California have destroyed over four million acres in what has been the worst fire season yet. Once again, the rainy season started late. For a while, only I505 separated the McNamara's walnut farm Sierra Orchards from the 'LNU Lightning Complex Fire'. The Frasier Ranch made it through another drought year. With record wind speeds of 140mph, a 'Derecho' cut a swathe of destruction through the Midwest.

My thanks go to Kris Nichols, Kate Scow and Cynthia Daley for patiently explaining rather complex facts and for being so passionate about soil. By being so vocal about the enormous potential of good soil and helping farmers and ranchers to 'climate proof' their operations by improving soil health they demonstrate how large scale climate mitigation can work.

There would not be a book without Petra Loch and Kambiz Ghawami, Fiona MacDonald and Mike Gardner. Thank you for making it happen.

And, from the bottom of my heart, I thank Martin Kunz – husband, driver, photographer.... During our research trips Martin took thousands of photos, some of them are included in this book (more can be found on LondonCowGirl.com). He transformed my schedule for interviews and farm visits in eight US states into an actual travel agenda. He got us everywhere – by plane, train, Greyhound bus and by driving several thousand miles. And he always found us a place to sleep at – though occasionally being really tired and not losing one's sense of humour did help. But more important than that, Martin stood by me through all the emotional ups and downs of this whole project. Without him and his love this book wouldn't exist.

Marianne Landzettel, London, November 2020

Marianne Landzettel is a journalist writing and blogging about food, farming and agricultural policies in the UK, continental Europe and the US. She was UK and Ireland correspondent for German Public Radio where she started her career as a reporter for the farming program. She worked for the BBC World Service from 2003 until 2013. Since then she writes and blogs for farming publications and other media outlets in the UK, the US and in Germany. She lives in London.
Twitter: @M_Landzettel
Internet: LondonCowGirl.com

Martin Kunz is a political scientist, Fair Trade pioneer, expert for setting up Fair Trade supply chains and photographer – and Marianne's husband. He deals with travel logistics – from plane, bus and train tickets to car hire and (occasionally exotic) sleeping arrangements and hotel bookings.

Index

A

Absdorf 230
agrochemical companies xiv, 48 - 50, 53, 124, 233
agroforestry xi, 85, 223 - 224, 228
ahupua'a 85
Alexander, Sophie 91, 232 - 233
alfalfa 21, 70, 74, 116, 121, 152, 156, 192, 194, 230, 237
Allentown, PA 18 - 19, 22
all natural 116 - 117
Almena, KS 177, 180, 184
Aloha Aina 63 - 64, 77 - 79, 89, 91, 99, 104
Amish 17 - 19, 33 - 35
AMMP, Alternative Manure Management Program 239 - 240
anaerobic systems 12
Anderson, Fiore 97 - 99
Angus 110, 119, 197 - 199
Animal welfare 251
annuals 9, 157, 190, 196
antioxidants 16, 250
Arkansas 43, 105, 167 - 170, 173, 210
Atay, Alika 62 - 65, 77 - 78, 104
atrazine 57, 70
Aubrac 199

B

BASF 48 - 49, 58, 93, 168
Batten, Lynn 102
Bayer-Monsanto 17, 48 - 49, 80 - 81, 168, 171, 233
beans 29, 98, 106, 159, 162, 164 - 168, 173 - 174, 187, 191 - 192, 212, 215, 230
Beaver City, NE 136
beef xiii, 9, 119 - 120, 123, 125, 136 - 137, 139, 146, 155, 160, 162, 179, 197, 199, 251
beef finishing 116 - 117, 137, 141
beneficial insects 3, 105, 159, 192, 233, 243
Bennett, Hugh 45 - 46
Berns, Keith 164, 175, 187 - 193, 199
Bethlehem Steel 18 - 19
Big Island 49, 72 - 75, 104, 107 - 108
biochar 106, 208, 237
biodynamic 24
Bird City, KS 130
Black Sunday 44, 46
Bladen, NE 186 - 188

blizzard 42, 44, 46, 154, 183
blue corn 197
blue grama grass 109
Boudreaux, MiKey 109
Bradley, Kevin 168, 170
Briggs, Stephen 223 - 228, 232
British White cattle 155, 162
Brown, Gabe vi, 154 - 155, 158, 164, 179 - 180, 189
buckwheat 175, 181, 191, 193, 230
buffalo grass 109, 118
buffalos 10, 12, 116
buffer zone 57, 61, 78
Butz, Earl 133

C

CalCan 239 - 240
calcium 2, 28, 181
California State Board of Food and Agriculture 243 - 244
Callicrate, Mike 137 - 141
Cameron, Don 243 - 245
Camp Fire vii, xvii, 210, 244 - 245
cancer 51, 56, 80, 98, 142, 184
Carbon viii, xiii, 108, 155, 242 - 243
carbon sequestration xi, xii, 11 - 17, 21, 95, 107 - 108, 198, 238 - 242
CDFA, California Department of Food and Agriculture 201, 239
Center for Food Safety, CFS 53
center pivot ix, 47, 138, 189
Central Valley xvi, xvii, 184, 201 - 203, 206, 210 - 211, 235, 237, 239, 243 - 244
cereals 9 - 10, 14, 25, 155, 185
ChemChina 17, 48, 52 - 53
chemical fertilizer xiii, 3, 13, 17, 25, 27, 71 - 72, 104, 125, 133, 175, 197
Cheyenne County, KS 128 - 130, 147, 149
Chico vii, viii, xii, xvii, 201, 210, 244 - 248, 250
chicory 192
Chico State University 246
chlorpyrifos 54 - 56, 78 - 79
Civilian Conservation Corps 45
climate change v, vi, vii, viii, ix, xvi, xvii, 8 - 14, 17, 29, 77 - 79, 94 - 96, 104, 106, 114, 156 - 157, 207, 214, 239 - 240, 243, 250
climate mitigation 198, 222, 255
cold season grasses 115, 158, 190, 196

Colorado vii, xvi, 37, 107, 109, 112, 115, 117, 127 - 128, 137 - 140, 144 - 146, 149, 184, 189
Colorado Springs, CO 137, 140
commodity prices 30, 133
communities, rural xiv, 17, 34, 42, 45, 58, 97, 100, 117, 122, 124, 149, 161 - 162, 166, 197, 203, 246
compost tea 207, 232 - 233
Consumers xiii, 76, 220
conventional farming vi, 16, 17, 20 - 22, 72, 74, 116, 123, 125, 132 - 136, 141 -142, 146, 159, 168, 190 - 191, 198, 212 - 213, 227, 233, 237 - 238
conversion 133, 135, 235, 253
copper 2, 46
Cornell University 24, 30 - 31
Corteva 48
cover crops xvi, 8, 13, 21 - 28, 30, 69, 123 - 125, 134, 141, 146, 148, 152 - 155 - 159, 163 - 164, 172 - 177, 179 - 181, 185- 198, 205, 207 - 212, 221, 227, 230 - 231, 237, 243 - 244, 247, 251
cow-calf producers 116
Cow peas 135
Coy, Richard 170
Crews, Tim 26
crop insurance xiii, 132, 181 - 182, 235
crossbreed 197
CSA scheme 215

D
Daley, Cynthia vii, xii, xvii, 245 - 248, 250, 252, 255
Devon, UK 217 - 219, 221
Dicamba 56, 131, 165 - 166
Dinnie Stone 121
direct seeding 22, 157
DowAgro-Science 49
Downsizing 124
drift damage 154, 166 - 167
drought vi, viii, xvi, 8, 21, 26, 30, 39, 42, 47, 48, 110 - 114, 115, 130, 135, 147, 154, 157, 159, 184, 188, 203 - 204, 213, 218, 227, 249, 255
dug-out 118, 129
DuPont 17, 48 - 49, 55, 67, 93, 168
Durum wheat 134
Dust Bowl 37 - 39, 42 - 48, 112 - 123, 130, 147, 184, 186, 211
dust storms 37 - 44, 45, 48, 112, 123, 130, 147, 184, 211

E
earthworms xiii, 31, 71, 183, 196, 230 - 233, 253
ecosystem 3, 6 - 8, 24, 85, 117 - 118, 155, 158, 175, 193, 198, 207, 242, 245
Egan, 38, 40 - 41, 46
Egan, Timothy 38, 40 - 41, 46
electric fences 179
Emporia, KS 151, 156 - 161
EPA (Environmental Protection Agency) 54, 65, 81, 168 - 171

F
fallowing 39, 44 - 45, 69, 100, 103, 106, 124, 130 - 131, 134, 180, 184, 204, 238
farm bill xiii, 235
FDA (US Food and Drug Administration) 76
feedlot 116, 123, 152, 182, 194
Ficke, Del 194 - 199
field trials 18 - 21, 66 - 70, 192, 195, 213, 233, 248
fly-over country 118
fodder crops 22, 25, 27, 30, 124, 158, 191, 195 - 196, 230
food safety 76 - 77, 244
Food Sleuth podcast 23, 170
food waste 216, 221
Frasier, Mark 109 - 122, 127, 156, 255
Fresh Seven Café 127, 137
fruit trees 144, 147 - 148, 223, 228
Fukushima 61
Fuller, Gail 151 - 161, 171, 174, 179 - 180
Fulton, Kevin 118 - 126, 129, 136, 141, 156, 162
fungi xiii, 3, 7 - 10, 207 - 208, 227, 233, 247, 250

G
Galloway cattle 119, 126, 144
GE corn 58, 71, 171, 182
General Mills 27
genetically engineered (GE) xiii, xiv, 48, 58, 63, 66, 70 - 79
GE plants 71 - 72
GE soy 48, 164, 167, 171, 173
GHG emissions xi
glomalin 6 - 9
glycoprotein 7
glyphosate 28, 56, 72, 80 - 81, 131, 141, 152, 154, 166 - 167, 171 - 173, 233, 252 -253
GMOs xiii, 58, 73, 76, 79, 142
golden rod 28
Grand 45, 252

Grand, Alfred 229, 230 - 234
Grapes of Wrath, John Steinbeck 37
grassfed 120, 125, 162
GrazeMaster 198 - 199
Great Depression 38, 112, 121
Great Mahele 91 - 92
Great Plains ix, 38, 45 - 46, 111, 147, 195
Green Cover Seed 187 - 193, 199
Green New Deal xi

H
HAACP 159
Haoles 60, 77 - 78, 91, 106
Hartung Brothers 52 - 53, 55, 68
Hawai'i xiv, 19, 48 - 108, 184, 235
Hawai'i Alliance for Progressive Action 77
Hawaii Crop Improvement Association, HCIA 68
Hawa'i senate bill SB30095 78
health insurance 15
health problems xiv, 60, 75
Hemmelgarn, Melinda 23 - 24, 170
herbicide resistance xii, 226
herbicides xii, xiii, xiv, 13, 17 - 18, 22 - 24, 56, 70 - 72, 93, 121, 123 - 125, 131 - 133, 154, 166 - 168, 172 - 175, 179 - 181, 184, 192, 198, 207, 211, 213, 226, 235 -237, 246, 251 - 253
Herefords 119, 194, 197, 199
high fructose corn syrup HFCS 58
High Plains Food Coop 142, 144, 148 - 149
Ho`oulu `Aina, Kalihi Valley Nature Reserve 91, 97 - 99
holistic grazing 112, 119 - 120, 126
Homestead Act ix, 129, 189
Honolulu 49, 56, 62, 68 - 69, 73, 78, 83, 86 - 89, 93 - 94, 96 - 99, 104, 107
Hooser, Gary 55 - 58, 66, 75, 77, 81
hybrid seeds 23, 70 - 72
hyphae 7, 9

I
Indigo Ag 241 - 242
Industrial agriculture 49, 251
iron 2, 6, 19, 110, 117, 186, 188
Isley, Jeter 143 - 149, 159, 221

K
Kahului 62
Kalihi Valley, Roots Café 86 - 87, 89 - 99
kalo 84, 86
Kanehaiua, Hiwa 100
Kansas Land Institute 25 - 27
Kauai 49 - 60, 62, 65 - 67, 72 - 75, 95, 105, 183
Kekaha Agriculture Association, KAA 93

Kernza 25 - 28
Kihei 61 - 63, 66
Kimbrell, George 73 - 79
KKV, community and health center, Kalihi 86 - 91, 97 - 98, 104
Klie, Joanne & Robert 136, 140 - 148
Krebsbach, Laura 138 - 141
K-State, Kansas State University 121, 134, 165
Kutztown, PA 18

L
Lake Seneca 23
Lakeview Organic Grain 32
Lancaster County 32
Last Chance, CO 32, 109, 127
Lee, Chris 94 - 96, 104
Lincoln, NE ix, 194, 199
loam 6, 184
longgrass prairie 118
Loup City, NE 121
Lukens, Ashely 58, 65, 73
Lundberg, Bryce xvii, 210 - 214
Lundberg Family Farm xvii

M
magnesium 2
Ma'o Farms 99 - 104
marestail 166, 173
marketing, direct x, 2, 18, 22, 49, 69, 77, 91, 95, 111, 125 - 126, 134, 141 - 142, 157, 159, 163, 174, 183 - 184, 192 - 193, 220, 222, 225, 239, 241
Martens, Klaas 22 - 35, 152
Martens, Mary Howell xv, 32, 34
Martin, Diana xiv, 20, 111, 118, 177, 186, 255 - 256
Maui 49, 58 - 80, 92, 95, 104, 107
Maupin, Marghee 50, 51 - 55, 77, 183 - 184
McNamara, Sean 205 - 210, 242 - 243, 255
McPherson, KS 136
meat chickens 160, 162
meatpacking 137
Mecklenburg-West Pomerania 39
Mennonites 34 - 35
Merrill, Jeanne 239 - 241
Meyer, Christian 252
Miles, Albie 92, 103 - 104
milo 157 - 158, 174
mobile henhouse 124
mobile slaughter 137 - 141
Moloa'a Organica 105 - 106
Molokai 49, 58 - 59, 64, 69

Monsanto 17, 48 - 49, 58 - 77, 80 - 81, 121, 124, 166 - 168, 171, 233, 242
mustard 25, 212, 219, 230
mycorrhiza 9, 71, 225, 253

N

National Research Conservation Service, NRCS 133
Nene geese 75
Ness, Autumn 60 - 62, 65 - 66, 77, 79
Nichols, Kris 1 - 17, 250, 255
nitrate 2, 13, 21, 106, 133, 148, 159, 163, 191 - 193, 230
nitrogen 25, 28, 30, 148, 187, 230, 239
non-GM varieties 21
no-till 21, 31, 131, 151 - 154, 164, 180, 194, 207, 230, 233, 247 - 248
nutrient density viii, 16, 250

O

Oahu 49, 85, 92, 94, 99 - 100, 104
Obama, Michelle 54, 103
Odom, Sharon Kaiulani 87 - 91, 97 - 98
Ogallala Aquifer 47, 188, 194
Omaha 183, 199
organic certification 20, 122, 125, 185, 209
Organic food 227
organic matter viii, 3, 12, 30 - 31, 44, 100, 124, 146, 148, 153, 157, 164, 180, 191 - 193, 197, 212, 227, 230 - 231, 237 - 240, 246 - 248
organic system 21, 208, 211
organophosphates 54
Osage 147

P

Pang, Lorrin 67 - 68, 71, 75 - 76
parent lines xiv, 69 - 72
Pawnee Nation 197
pearl millet 175
Pennsylvania xv, 18 - 21, 33 - 34, 168, 233
Penn Yan, NY 23, 26, 31 - 35
perennials 190
permanent grassland 9, 13, 46, 146, 156 - 158, 185
Peterborough, UK 219, 223, 228
phosphate 133, 152, 163
photosynthesis viii, 2, 3, 8, 10
pigweed 166, 173
pioneer 55, 135, 151, 256
plantain 192
Practical Farmers of Iowa, PFI 177

prairie xvi, 12, 25 - 26, 40 - 41, 45, 47 - 48, 109 - 118, 120 - 123, 156, 158, 190, 194, 196
Puerto Rico 79, 93, 95

R

radishes 56, 175, 180, 191, 196
ragweed 28, 152, 158
Raile, Robyn 129 -130, 136
Raile, Tim 26, 118, 127 - 136, 141, 146, 158
rainfall ix, xiii, xvi, 14, 16, 30, 41, 47 - 48, 111, 113 - 114, 116, 118, 122, 134, 157, 184, 192 - 193, 203, 227, 245
reducing costs 130
Red Winter Wheat 129, 133
reefer 139 - 140
regenerative agriculture vi - xi, xiv, xvii - xviii, 17, 20, 122, 130, 151 - 154, 160 - 161, 179 - 180, 185 - 186, 235, 239 - 241, 247 - 254
Regenerative Agriculture Initiative vii, 248, 250
Regenerative Organic Certification (ROC) 20
Restricted Use Pesticides, RUPs 56
Ritte, Walter 58 - 60, 78
Riverford 215, 217 - 222
Rocky Mountains 40, 111
Rodale Institute xv, 1, 18 - 19, 25, 233
roller crimper 22
Roosevelt, Franklin D. 45 - 47
rotation 21, 97, 100, 106 - 107, 130 - 131, 152, 163, 181, 195, 211 - 213, 221, 226, 230, 234, 237 - 238
Roundup Ready Xtend 167 - 168
ruminants 9, 10, 198
run off 8, 157
Russell Ranch 236 - 240
Russian-Germans 40
Ryals, Rebecca 107 - 108

S

Sacramento, CA 206, 210 - 212, 236, 239, 242
San Joaquin Valley 203, 210, 243 - 244
Savory, Allan 111 - 112
Scow, Kate 236 - 239, 255
SEED 57
seed corn 81
seed mix 175, 190 - 191
sharecropping 133
shelterbelt 147
shortgrass prairie 116, 118
Sierra Orchards 206 - 207, 210, 243, 255
silt 6, 30 - 31, 107
silvoarable 223 - 225, 228
Simmental 110
simple sugars 2, 7, 10

slaughter 116 - 117, 120, 126, 136 - 141, 156, 159, 184, 252
slaughter weight 116 - 117, 120, 126, 156
slurry 13
Smith, Pete 11 - 16, 43
soil aggregate 7
soil biology 1 - 2, 11, 71, 246 - 247
soil compaction 191, 207
Soil Conservation Act 46
soil degradation 45, 154
Soil Health initiative xi
soil moisture 12, 130 - 131, 148, 171, 174, 181, 188 - 189, 193, 204
soil organic matter viii, 3, 30, 44, 100, 124, 148, 153, 164, 180, 191 - 193, 197, 227, 237 - 240, 246 - 248
soil organisms 1 - 3, 6, 8, 13, 16, 31, 34, 107, 151, 196, 198, 209, 230, 238, 241, 246
soil samples 107 - 108, 229, 233, 247
SOM, soil organic matter viii, 3, 30, 44, 100, 124, 148, 153, 164, 180, 191 - 193, 197, 227, 237 - 240, 246 - 248
sorghum ix, 156 - 158, 175, 180 - 181, 188, 191, 194 - 196
spam 90
spraying x, 24, 53, 56, 58, 60, 65, 71, 73, 168 - 169, 179
spray table 66
stacking enterprises 228
Starke, Michelle 69 - 70
Steinbrecher, Ricarda 71 -72
St. Francis, KS 127 - 130, 134 - 137, 142, 144 - 145, 158, 185
subsoil 30, 44, 123, 135, 152 - 153, 180
sudangrass 175, 180 - 181, 191
sugar cane 53, 58 - 66, 91 - 93, 96, 105
suitcase farmers 40 - 41
sulfur 2, 28, 239
Sunhemp 191
supermarkets 58, 103, 162, 202, 216 - 217, 220, 252
superweeds 17
Svobida, Lawrence 42 - 45, 186
SWEEP, State Water Efficiency and Enhancement Program 239
switch grasses 118
Syngenta 17, 48 - 49, 52 - 55, 68, 75 - 76, 93, 121

T
tap roots 147, 191, 230
taro 78, 83 - 87, 90, 97 - 98, 106
Terkel, Studs 38
test plot 27
Thompson, Michael 42, 118, 177 - 186
Three Mile Island 18
tilling 131, 141, 178
tilth 44, 107, 157, 196, 227, 247
topsoil xvi, 8, 30, 44, 123, 135, 142, 153, 175, 196 - 197
Turkey Red wheat 158
turkeys 126, 161 - 162

U
UC Davis, CA 236, 238, 240
UN Climate Change Conference x
Underseeding 25
University of Hawai'i at Manoa 71
University of Hawai'i, West O'ahu 92
USDA 20, 65 - 66, 70, 79, 81, 138 - 140, 147, 171

V
Valenzuela, Hector 71, 96 - 97
vegan 10, 221, 251, 253
vetches 191
volatility 167 - 169, 173

W
Waimea 51 - 55
Wall Street Crash 38
walnut trees 209
warm season grasses 115, 158, 196
waterhemp xii - xiii, 166
water holding capacity 148
water infiltration 8, 111, 114, 142, 148, 157, 180, 230 - 231, 247
Watson, Guy 217 - 222
Waverly, KS 161 - 163, 175
Wegemans 35
Whitehall Farm 223, 228
Whitlock 105 - 106
Wilhelm, Dean 83 - 86, 90, 98
Williams, Darin & Nancy 161 - 166, 171 - 175
Winters, CA 206, 242
worm castings 231 - 232
worm compost 106, 229 - 232

Y
Y Knot Farm 144 - 148

Farming with benefits

Ingram Content Group UK Ltd.
Milton Keynes UK
UKHW051258280623
424193UK00007B/29